A Special-Interest Car Buyer's Guide

MARQUES of AMERICA

Edited by John Gunnell

D1403770

PHOTO CREDITS

© 1994 by
Krause Publications, Inc.

krause publications
since 1952

700 E. State Street • Iola, WI 54990-0001
Telephone: 715/445-2214

Library of Congress Catalog Number: 94-75300
ISBN: 0-87341-302-4
Printed in the United States of America

Contents

Introduction 4
AMC Family Highlights 6
Buick Highlights 7
Cadillac Highlights 8
Checker Highlights 10
Chevrolet Highlights 12
Chrysler Highlights 14
Crosley Highlights 16
DeSoto Highlights 17
Dodge Highlights 19
Ford Highlights 21
Hudson Highlights 23
Imperial Highlights 25
King Midget Highlights 27
Lincoln Highlights 29
Mercury Highlights 31
SHADES OF THE PAST - Full-Color Section 40
Oldsmobile Highlights 41
Packard Highlights 43
Plymouth Highlights 45
Pontiac Highlights 46
Shelby Highlights 48
Studebaker Highlights 49
Tucker Highlights 51
Special-Interest Car Review 53
1968-1969 AMX: built for speed 54
1936 Buick 8 "as positive as steel on steel" 57
1940 Buick: new features and modern styling 61
Buick's image changed in 1953 64
This Century was a Special 68
1959 Buick Indy Pace Car 70
1963: The premier Buick Riviera 73
Cadillac's last V-16 76
A Cadillac coup: the 1949 DeVille 79
Cadillac's last 1960s ragtop 82
The meaning of Marathon 85
Expect wonderful things from 1953 Chevrolets 88
Beautiful basics: 1958 Chevrolet Biscayne 92
Down-sized 1961 Chevrolet 97
Corvette introduces the Sting Ray 100
1968 Caprice: Chevrolet luxury 103
1968 Nova Super Sport 107
1973 Nova "hatched" new idea 110
The car that swept Daytona:
    1955 Chrysler 300 113
Chrysler 300B took 16 wins in a row 116
The world's most complete car? 119
1949 Crosley: Hotshot prediction 124

DeSoto Adventurer debuted in 1956 127
Last DeSoto-built Adventurer 130
1959 brought DeSoto lots of "lasts" 132
Dodge dynamite: the D-500 136
MoPar's early muscle: 1962 Dodge 141
Charger: Dodge's first-class fastback 145
1939 Ford ragtop is oldest, yet latest thing 148
1940 Ford Deluxe was handsomely new 151
Ford had a high-style sedan in 1950 154
Ford's first T-bird: 1955 158
1958 Ford Custom 300 161
1960 Fords, the forgotten collectible 163
Ford's 1961 Country Squire 166
1961 Thunderbird unique in all the world 169
Hudson's Country Club for 1939 172
1954 Imperial: Eisenhower era exclusive 174
1961 Imperial: The car that time forgot 176
1962 Imperial challenged the
    luxury car market 179
1954 King Midget: little, but long-lasting 184
Lincoln's new "teardrop" ragtop for '38 186
1951's "small" Lincoln 189
$10,000 without air: the 1956 Continental Mark II 192
Mark III embodied personal luxury 195
1949 Mercury woodie was different 198
The last of Mercury's Marauders: 1969-1970 201
Oldsmobile's B-44 203
1958 Oldsmobile: mid-price sales leader 207
Luxury in a large package:
    1973 Oldsmobile 98 211
New management pushed 1953
    Packard Patrician 214
Zig-zag Packard 217
1957 Packard Clipper Country Sedan 221
The 1950 Plymouth 224
1956 brought Plymouth's first muscle car 227
1969 Road Runner: no-frills super car 229
1937 Pontiac was a great car for a great year 233
Big time 1941 Pontiac coupe 236
1963 Grand Prix: plain was beautiful 239
Ordering a 1969-1/2 Firebird Trans Am 244
1962: When the Cobra struck 247
Styling steadied Studebaker 250
Studebaker was first in '55 253
Rare, radical 1948 Tucker 256
On your marque 259
American Marque Car Clubs 260

INTRODUCTION
Marques of America

The computer that this book was typed into has a spell-checker function. Strangely enough, its spell-checking dictionary does not recognize the word "marque." Nevertheless, this word is instantly recognizable to car collectors, who use it often to make reference to a make or brand of automobiles.

Some people pronounce or write the word as "mark" or "marquee." Both would seem improper. *Webster's New World Dictionary*, Third College Edition, 1988 indicates that marque (pronounced mark) refers to the token of a pledge and is an obsolete word, except in Letters of Marque. However, it gives a second set of definitions as: 1) a nameplate or emblem to identify an automobile; 2) a brand name of a product, especially of an automobile.

Apparently, marque does have a relationship to a mark (as used to denote a sign written on a pledge in place of a signature), but has nothing to do with a marquee, which is a word that describes a type of canopy or tent and is now commonly used to describe theater entrance structures.

Marques of America, then, is a book about cars with automotive brands and nameplates indigenous to the United States. The sub-title *A Special-interest Car Buyer's Guide* suggests that we're talking about particular types of cars produced from about the mid-1930s to the mid-1970s.

"Special-interest cars are cars that are out of the ordinary and of unusual or special interest to automotive enthusiasts. The most important element is that they are not ordinary." This definition of the term appeared in the 1956 edition of *Classic Cars and Specials* by Robert J. Gottlieb, the well-known classic car editor for *Motor Trend*.

"Because the field is new, it is difficult to appraise a given special-interest car with that degree of assurance with which a known classic car can be appraised," Gottlieb wrote 38 years ago. He went on to suggest that classics were generally more valuable than special-interest cars, but noted that models like a Tucker (worth $3,000 then and $300,000 now) could be exceptions.

The definition of a special-interest car hasn't changed essentially since 1956. However, the number of cars that fit into this category has grown.

Today, a special-interest car can be anything from a 1935 Studebaker President sedan to a 1973 Chevrolet Nova SS hatchback.

The first section of this book will run down historical highlights for each of 21 companies that produced the special-interest models featured inside. It tells about the cars of each marque that are reviewed in detail. Other "Picks & Prices" are also charted for each brand. The charts list 10 additional models of each nameplate which are of special interest to marque collectors. They also give the *Old Cars Price Guide* number 1 condition price for each of the 10 cars picked. These are expert "ballpark" estimates of what each car, in top shape, is worth in today's collector car market.

Marques of America has two additional sections. The "Special-Interest Car Review" section takes a close look at the 21 American marques, giving details on particular models. There are 65 entries on particular cars with photos of the featured model or related models and a specifications chart giving the vital statistics. Prices are included once again.

Finally, the "On Your Marque" section lists clubs and other information related to countless other American marques which you might encounter in your treasure hunts for cars, parts, literature or automobilia.

John Gunnell
March, 1994

Marques of America

AMC Family Highlights

American Motors Corporation (AMC) traces its history back to early automobiling at the turn-of-the century. It was Jeffery then. Charles Nash was among the purchasers of the firm in 1916. Nash automobiles were built 1917 to 1957. However, the AMC identity emerged in May 1954 when Nash and Hudson joined forces. AMC wasn't used as the name of a product, though, until the mid-1960s. Then a new red, white and blue logo began to appear in advertisements.

Automotive historian John Conde, who worked in AMC's public relations department, produced a book called *The AMC Family Album*. This book became the "bible" for collectors of all brands of vehicles related to the AMC marque, including Jeffery, Nash, Rambler, Hudson, Essex, Terraplane, LaFayette, Dover, Willys, Jeep, Nash-Healey, Metropolitan, AMC and Eagle. Therefore, it's common to hear enthusiasts talking about the "AMC Family."

Representing AMC in this book is the very collectible AMX two-seater of 1968. It was designed to give AMC a more youthful image and earn it a share of the growing muscle car market.

AMC Family Picks & Prices

(1) 1940 Nash "600"..$7,800
(2) 1946 Nash Suburban..$23,000
(3) 1948 Nash Ambassador Custom Convertible...$25,000
(4) 1950 Rambler Landau Convertible $6,500
(5) 1957 Rambler Rebel $9,000
(6) 1961 Rambler American Convertible..$6,200
(7) 1966 Marlin $5,500
(8) 1966 Rambler Rebel Rogue $4,500
(9) 1969 AMC SC/Rambler..$13,000
(10) 1970 AMC Rebel Machine..$12,500

The Rambler Rebel Rogue was one of AMC's first hot, V-8 powered, youth-market cars of the 1960s. It had a special midyear introduction at the Chicago Automobile Show. (American Motors)

Buick Highlights

Buick was organized in 1903, when 16 cars were built. In 1906 came the first four-cylinder Buick with sliding gear transmission. The company joined General Motors in 1908, thereby becoming a cornerstone of America's foremost automaker. The six-cylinder Buick was introduced in 1916. By 1927, the Flint, Michigan company had built 2,000,000 cars. The low-priced series 40, which later used nameplates like Special, LeSabre and Invicta for added identity, evolved for 1934.

Early in the special-interest era, the 1938 Buick became America's third most popular auto marque. By 1940, Buick had built its four millionth car. War materiel manufacture began in 1941. After the war, Buick introduced Dynaflow in 1948 and hardtop styling in 1949. By the early 1950s, Buick had developed a conservative image that had an upside and a downside. The 1953 Skylark was developed as a limited-edition 50th anniversary model designed to give the firm more pizzazz. Still, Buick's rank in industry slid during the late 1950s.

The compact new Buick Special, for 1961, previewed an age of new thinking and marketing experimentation. Among the ideas tried were V-6 engines. A stylish 1963 Riviera announced an effort to tap into the personal/luxury car market. Buick also entered the muscle car realm with GS and GSX models and Stage I/Stage II engine options.

Representing Buick in this book are models from the 1930s, 1940s, 1950s and 1960s.

Buick Picks & Prices

(1) 1948 Roadmaster Convertible	$29,000
(2) 1949 Roadmaster Convertible	$32,000
(3) 1954 Skylark Convertible	$45,000
(4) 1955 Century Riviera Sedan	$11,500
(5) 1958 Limited Convertible	$39,000
(6) 1962 Buick Wildcat Sport Coupe	$10,500
(7) 1964 Custom Sport Wagon	$6,600
(8) 1966 Skylark Gran Sport Convertible	$15,000
(9) 1970 GSX Convertible	$17,000
(10) 1971 Riviera GS Boattail Coupe	$6,500

In 1954, the Skylark Sport Convertible remained a limited-edition model, but it was transferred from the Roadmaster platform to the Century chassis. The second Skylarks were rarer than first-year models and only 836 were made.

Cadillac Highlights

The first car from Cadillac Automobile Company was built in 1902. Cadillac became part of General Motors in 1907. A closed body was standard equipment, which was unusual at that time. Cadillac introduced the first V-8 engine used in a mass-produced car in 1914. The company moved into its factory on Clark Avenue, in Detroit, Michigan, during 1919.

As the special-interest car era started, Cadillac pioneered the integral casting of V-8 engine blocks and crankcases for 1936. In 1942, civilian car production was halted so that the automaker could build M-5 tanks and M-8 motorized gun carriages.

Big news came from Cadillac, after the war, in the form of a new overhead valve V-8 engine and a hardtop that looked like a convertible called the Coupe de Ville. By the time 1950 had arrived, the company was realizing all-time production records and adding workers for 1951. That year marked Cadillac's highest employment in peacetime history. A restyling in 1952 established design leadership for the entire industry, a position that Cadillac maintained into the 1960s. Innovations for the division included bumper-exhaust outlets, dagmars, tailfins, narrow band whitewall tires and convenience options ranging from ladies' make-up cases to sensors that raised convertible tops when it rained outside.

Standard Cadillacs grew more conservative-looking in the 1960s, but established domination of the luxury car market. A series of ads from the era even depicted late-model used Cadillacs with their showroom-fresh counterparts to emphasize high re-sale values. For the younger-at-heart with a taste for adventure, the front-wheel-drive Eldorado coupe made the scene in 1967. It was restyled in 1971, bringing back a taste of the 1950s with its rounded Coke-bottle shape and skirted rear fenders.

Cadillacs featured inside are 1940, 1949 and 1969 models.

Cadillac Picks & Prices

(1) 1938 60 Special Sedan $35,000
(2) 1941 Series 62 Convertible $51,000
(3) 1947 Series 61 Four-door Fastback $18,000
(4) 1948 Series 75 Business Imperial Limousine $25,000
(5) 1953 Coupe de Ville $26,000
(6) 1956 Series 62 Convertible $42,000
(7) 1957 Eldorado Brougham $29,000
(8) 1962 Series 62 Convertible $30,000
(9) 1970 Eldorado Coupe $11,000
(10) 1973 Coupe de Ville $6,500

A double-drop frame permitted the 1938 Sixty Special to sit three inches closer to the ground than other Cadillacs of the same vintage. Its new body, had no runningboards and featured large, flush side windows. (Cadillac)

The 1953 Eldorado convertible was nicely styled. It sported the year's redesigned grille with a heavier, integral bumper. The famous Cadillac ragtop was priced just under $8,000 and only 532 were sold.

Checker Highlights

Evolving out of companies based in Buffalo, New York and Detroit, Michigan, the Checker Cab Manufacturing Company was founded in 1922 and situated in Kalamazoo, Michigan. While the company focused on making taxicabs, there were at least two periods of making passenger cars. The first such product was an interesting all-steel "suburban" or station wagon. Thus, while Chevrolet proudly claims to have introduced this body style, Checker can actually take credit. At the start of the special-interest era, Checker was producing purpose-built New York and Chicago taxis, with unusual open-landau roofs. A few custom-body designs and armored cars were made, too.

Checker took a stab at developing a prototype for the U.S. Army Jeep in 1940. The all-wheel-steer experimental was rejected due to expense. After all, it was hard to build a cheap Checker. In the postwar 1940s and 1950s, Checker sold thousands of look-alike cabs that somewhat resembled a Chrysler product, as far as styling went. These taxis starred in many childrens' books in the bright yellow-and-green or yellow-and-red colors. Few, if any, of these models survive today. In 1958, the firm changed its name to Checker Motors Corporation.

Late in the special-interest era, Checker returned to marketing cars to private owners. This began around 1960. The cars looked just like the taxis and utilized the same Continental-built L-head six-cylinder engine. There were only minor decorative and trim differences between the commercial and non-commercial versions. In addition to four-door sedan big enough to hold a horse in the rear (as depicted in one publicity photo), Checker made a longer wheelbase Marathon, wagons and airport stretch-limos based on the same body. There were constant improvements to power trains, features and options, safety equipment and dress-up items, but the 1960 Checker was basically the same as the 1973 model.

The 1962 Checker is detailed in this book.

Checker Picks & Prices

(1) 1936 Checker Model Y Taxi..$16,000
(2) 1939 Checker Model A New York Taxi...$16,000
(3) 1940 Checker Model A Landau Taxi..$16,000
(4) 1948 Checker Taxi..$13,000
(5) 1955 Checker Taxi..$9,500
(6) 1963 Checker Marathon Sedan...$7,600
(7) 1964 Checker Town Custom Limousine...$8,100
(8) 1965 Checker Marathon Station Wagon..$7,600
(9) 1966 Checker Marathon Town Custom..$8,000
(10) 1967 Checker Aerobus...$10,000

The 1931 Checker Model A Landau Taxi has a very distinctive body style and would be a good addition to a special-interest car collection. Where to locate one of these is a problem, though. (John A. Conde Collection)

Checker's version of the airport stretch limousine was an eight-door station wagon-like model called the Aerobus. It had a built-in luggage rack on the roof. This is the 1972 edition.

Chevrolet Highlights

Chevrolet Motor Car Company was organized in 1911 and introduced its first car the following year. The firm became best-known for its overhead valve engines, the first of which appeared in 1913, the same year the home plant was established in Flint, Michigan. Five years later, "Chevy" came under the General Motors umbrella. It was a good move for the corporation, as Chevrolet took the lead in American car sales away from Ford in 1927. By 1929, the famed "Stovebolt Six" was adopted as Chevrolet's power plant.

The special-interest car era started out with a lack of excitement at Chevrolet. In some years in the late-1930s, a convertible was not even offered. The 1940 Chevrolet's sharp appearance pumped the energy level upwards. By the time that the first postwar Chevrolet was made on October 3, 1945, plans were being made to bring Chevrolet up-to-date in the 1950s.

A change towards sportiness started with a wood trim kit released for early postwar two-doors. Though not factory made, the kit was sold through authorized Chevrolet dealers and greatly enhanced the cars' looks. By 1950, the Bel Air hardtop debuted. Next came the Corvette, a sports car with automatic transmission. It was followed by a hot, small-block, V-8 in 1955, tailfins in 1957 and a sporty Impala series in 1958. By 1959 and 1960, Chevrolet's image had totally changed.

Big-block Chevrolets drag racing, on weekends, gave the ever-affordable Chevy the youth-market following that all automakers craved. No one could deliver the same performance-per-dollar as good old "USA Number 1." Chevrolet spread out from its family car base to become all things for all of America. The sporty and radical Corvair compact, the bite-sized Nova and the intermediate Chevelle all helped broaden Chevrolet's marketing base. Another addition was the 1966 Caprice, which pushed the Impala towards a Cadillac-like image with Chevy-like price tag. To round out its lineup in 1967, Chevrolet launched the late-comer Camaro into pony car land.

Chevrolet recaptured some lost market share with the Monte Carlo sports/luxury coupe and the sub-compact Vega in the early 1970s. Also, some of the hottest-ever Chevrolets were issued in this era. They included the SS 454 and a budget bomb called the "Heavy-Chevy." The muscle car movement was dead, though, and the last special-interest era Chevrolets were full-size ragtops and wimpy Corvettes with dry-looking plastic rear bumpers.

In *Marques of America* we look at Bel Airs, Impalas, Caprices, Novas and Corvettes in our Chevrolet sampling. Some cars covered are valuable and rare "milestone" models, while others are treasures you can find and buy on a "crystal ball" basis.

Chevrolet Picks & Prices

(1) 1940 Special Deluxe Convertible $21,000
(2) 1948 Fleetmaster Station Wagon $25,000
(3) 1955 Bel Air Nomad Station Wagon $17,000
(4) 1957 Bel Air Convertible $35,000
(5) 1962 Corvette Convertible $33,000
(6) 1965 Chevelle Malibu Convertible $16,000
(7) 1967 Camaro SS 396 Convertible $19,600
(8) 1969 Nova SS-396 .... $8,500
(9) 1970 Monte Carlo SS-454 .... $14,500
(10) 1973 Caprice Classic Convertible $10,500

Chevrolet's 1940 Special Deluxe Convertible was a popular collector car even before the same company's mid-century models became collectible. It marked the beginning of a move towards increased sportiness in Chevrolet styling.

There is still a good market for Chevrolet convertibles of the early 1970s. Only 7,339 of the 1973 Caprice Classic convertibles were built. The last year for a really full-size Chevrolet ragtop was 1975. (Chevrolet)

Chrysler Highlights

On July 31, 1923 the first Chrysler came off the Maxwell assembly line in Detroit. The 1924 model had a 70-horse-power six and found 32,000 buyers. A four-cylinder car was added to the line on June 6, 1925. In 1930, the news was the introduction of a Chrysler straight eight. Floating Power was introduced in 1931, booster braking was added in 1933, the Airflow arrived in 1934 and 1936 brought assembly of the one millionth Chrysler.

Chrysler was among the first special-interest autos to have a fluid drive transmission, making it an option as early as 1938. For the postwar era, Chrysler added 4.5 million square-feet of factory space and geared up for a big future that came slowly. Engineering was long the company's forte and some of its late-1940s improvements included full-flow oil filters, Oilite fuel-tank filters, automatic back-up lights, water-proof ignition systems, cycle-bonded brake linings and pressure-vent radiator caps. All-electric window lifts were added to the option list for 1950 and the next year brought the "Firepower" hemi V-8, Hydraguide power steering and Oriflow shock absorbers.

Chrysler's "Forward Look" styling was thrust into the limelight in 1955. The low, long, wide look with tower-like taillights tucked in twin fins was aircraft-inspired and exciting. Quality was spotty, but Chrysler was a powerful force in the horsepower race. Innovations included the Chrysler 300 letter cars and the very first optional production car engine with one-horsepower-per-cubic-inch. Other firsts were Uni-body construction, torsion bar front suspension, optional record players and swivel-type driver seats. Chrysler seemed to be first in everything, except sales.

Early 1960s Chryslers looked like 1950s models that got a better job. They had "hair trims" and "classier duds" hung on the same basic bodies. This changed dramatically, in 1962, when the fins were literally "shaved" off the Chrysler body. Thereafter, the clean and sculptured look took over at MoPar headquarters. Part of the reason for the drastic change was house-cleaning following a management scandal that sent heads rolling in the corporate offices. When the dust had cleared, Chryslers had a pleasing high-tech luxury look.

The late 1960s and early 1970s saw a return to the fascination with aeronautical styling motifs. It was described as the "fuselage" look and involved a cigar-shaped profile mated to integrated bumper/grille styling. As impressive as the large size of these Chrysler bodies was the cubic inch figures for the power plants, which usually had 400- and 440-cubic-inch displacements.

In this book, we look at 1955, 1956 and 1963 Chryslers up close. Two letter cars are reviewed, plus a rare four-door version of the New Yorker.

Chrysler Picks & Prices

(1) 1935 Chrysler Airflow C-1 Sedan $20,000
(2) 1936 Chrysler Airstream C-8 Convertible Sedan $30,000
(3) 1939 Chrysler New Yorker Sedan $12,000
(4) 1941 Chrysler Town & Country Barrel-back $26,000
(5) 1951 Chrysler New Yorker Convertible $22,000
(6) 1958 Chrysler 300D Convertible (EFI) $75,000
(7) 1961 Chrysler 300F Convertible (four-speed) $62,500
(8) 1965 Chrysler 300L Convertible $2,000
(9) 1968 Chrysler 300 Hardtop Coupe $7,500
(10) 1970 Chrysler Hurst 300 Hardtop $10,000

What a shame to see a car like this 1936 Chrysler Airstream convertible sedan wrecked in a major collision. Only 362 examples of this model were ever manufactured. It sold for almost $1,300 when new.

Though not as rapid or rare as earlier 300 letter cars, this 1967 Chrysler 300 convertible sure is nice-looking.

Crosley Highlights

Powel Crosley, Jr. had many business ventures going from the manufacture of Crosley radios and refrigerators to ownership of a radio station and a professional baseball team. He tried to sell his first car in 1907 and marketed a cyclecar in 1911. Neither effort was a success. Later, after making millions in the appliance and entertainment fields, he introduced his longest-lasting car. The small, economical Crosley started production in a Richmond, Indiana factory during 1939. It survived through 1952.

During those 11 years, the car was offered in a full range of models (the high point was five distinct car body styles, in a single year), several trim levels (Standard and Super for 1950-1952), and in up to four separate series (Standard, Super, Hotshot and Super Sports in 1951 and Standard, Super, Hotshot and Super Shot in 1952). For commercial vehicle users, Pick-up, station wagon and panel delivery models were first added to the line in 1940.

In 1945, the company became Crosley Motors and was incorporated in Cincinnati, Ohio. A plant was purchased in Marion, Indiana for final assembly operations. By 1948, this factory underwent a 40 percent expansion and a service parts plant was purchased in Cincinnati. Production high-points were 1949, when some 18,300 were made, and 1950, when over 28,000 (including 23,849 station wagons) were built.

In historical and hobby terms, this makes the Crosley much more than just a "minor make." It spans the prewar and postwar eras, it represents significant total production, and it has a substantial enthusiast following, with most marque collectors holding membership in the Crosley Automobile Club, Incorporated.

Our "Special-Interest Car Review" features the 1949 Crosley Hot Shot.

Crosley Picks & Prices

(1) 1939 Convertible $4,000
(2) 1940 Pickup $3,400
(3) 1941 Parkway Delivery $3,600
(4) 1941 Covered Wagon $4,000
(5) 1946 Station Wagon $4,300
(6) 1949 Convertible $4,500
(7) 1950 Super Convertible $4,600
(8) 1951 Hot Shot $6,000
(9) 1951 Super Sports $6,300
(10) 1952 Super Shot $6,300

The small, economical Crosley looked like this when it came out in 1939. The regular models survived, with one major design change, through 1952. There was also an early 1950s sports car version, which is featured inside.

DeSoto Highlights

The DeSoto was added to Chrysler's lines in 1928. The first car was a 55-horsepower Six built in a new Plymouth plant. Over 200 cars per day were cranked out and first-year sales for 1929 totaled 81,085 units. At the time, that was an all-time record for any brand new automobile marque. Production was shifted to the Chrysler's factory at Jefferson Avenue, in Detroit, three years later. A plant on the city's Wyoming Avenue was purchased from General Motors, in 1934, to house production of DeSoto Airflows. All final assembly was moved there in 1936.

After making only Airflows in 1934, DeSoto entered the special-interest era building both Airflow and Airstream models. The latter held the company up during the depression, as did its entry into the taxicab market. Classic car designer Raymond Dietrich restyled the good-looking 1937 DeSotos and the last prewar model, offered in 1942, had hidden headlights.

Early postwar DeSotos were dusted-off 1942 editions, with conventional headlamps and toothy grilles, which would become a marque trademark through 1955. The one millionth DeSoto was manufactured in 1949, based on initially strong postwar sales. This business also inspired the opening of a new body plant on Warren Avenue in 1950 and its 1951 expansion to include engine manufacture. In the last year, DeSoto saw its third consecutive season of 100,000-plus sales.

DeSoto introduced the hemi V-8 in 1952 and marked its 25th anniversary the next year. Then came 1954 and a dramatic slide in sales. Designer Virgil Exner's "Forward Look" of 1955 led to an 85 percent sales increase and DeSoto had a new image to go with its high-performance engine. The "bubble" didn't last. By 1958, DeSoto entered a perplexing and permanent downtrend that ended after just over 3,000 last-of-the-marque 1961 models were built. The date was November 30, 1960 and the DeSoto was no more.

Marques of America: A Special-interest Car Buyer's Guide takes a close-up look at three exciting 1950s DeSotos. It reviews the 1956 Adventurer Sportsman hardtop, the 1958 Adventurer convertible and the 1959 Firesweep Sportsman in detail.

DeSoto Picks & Prices

(1) 1936 S-1 Custom Convertible Sedan $26,000
(2) 1942 S-8 Convertible $26,000
(3) 1948 S-13 Custom Suburban $7,800
(4) 1949 Custom Convertible $18,000
(5) 1955 FireFlite Convertible $26,000
(6) 1957 FireFlite Adventurer Sportsman $22,000
(7) 1958 FireFlite Adventurer Convertible $37,000
(8) 1959 Adventurer Convertible $27,000
(9) 1960 Adventurer Sportsman Coupe $10,500
(10) 1961 FireFlite two-door hardtop $11,000

The last DeSoto before World War II was the 1942 model with its unique hidden headlights. Although some low-volume cars such as Grahams and Cords had this feature, DeSoto was the first mass-producer to offer it. (Chrysler Historical)

Strong postwar sales led to a new body plant in Detroit. In 1951, the plant was expanded for engine manufacture. For DeSotos, including this 1951 Custom sedan, it was the third year for 100,000-plus sales. (Chrysler Historical)

Dodge Highlights

It was way back in 1901 that the Dodge brothers formed a partnership to produce automotive parts. In 1914, Dodge Brothers Incorporated began production of a four-cylinder touring car. An all-steel sedan was added to the line in 1918. During 1920, both of the brothers passed away. Only three years later, the one millionth Dodge was produced.

In 1925, Dillon, Reed purchased Dodge as an investment. Chrysler, seeking a low-priced line to sell, took ownership of Dodge on June 1, 1928. Within a year, the two and one-half millionth Dodge was turned out. By 1934, Dodge was America's fourth largest automaker.

Dodge entered the special-interest era as a volume producer of lower-medium-priced family cars. Playing off the strength of its dependable L-head sixes and eights and its strong all-steel bodies, Dodge built its prewar reputation on engineering and sparkling, though somewhat conservative, styling. The company marked its 25th anniversary in 1939. Early postwar Dodges stayed much the same. It didn't hurt. The five millionth Dodge was assembled in 1946.

Postwar changes to the line were made in 1949, when Coronets, Meadowbrooks and Wayfarers replaced the old Standard, Deluxe and Super Deluxe models. There were new all-steel station wagon and roadster models, too. Around 1954, Dodge began moving in the direction of affordable high-performance beginning with an Indy Pace Car and ending with the D-500/D-501 options. The performance theme continued in the 1960s and early 1970s, especially after the company's introduction of compact and mid-size models like the Lancer, Dart, Coronet and Charger.

In this book, the 1956 Dodge D-500, 1962 Dodge Dart and 1966 Dodge Charger are reviewed.

Dodge Picks & Prices

(1) 1935 DU Rumbleseat Convertible $27,000
(2) 1949 Wayfarer Roadster $21,000
(3) 1950 Diplomat Hardtop Coupe $11,000
(4) 1954 Royal Indy Pace Car Convertible $25,000
(5) 1959 Custom Royal Convertible $30,000
(6) 1963 Dart GT Convertible $9,000
(7) 1965 Coronet 500 Convertible $11,000
(8) 1967 Monaco 500 Hardtop Coupe $7,500
(9) 1969 Charger Daytona $59,000
(10) 1970 Coronet Super Bee $20,000

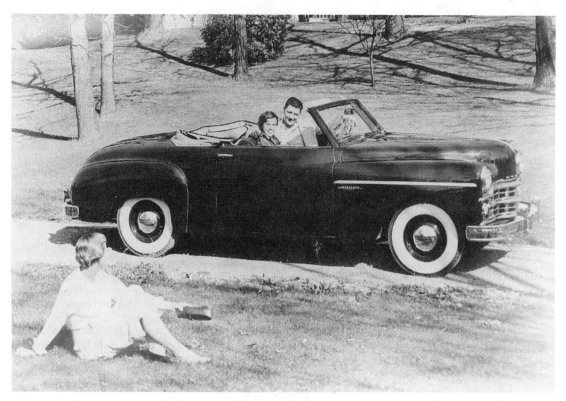

Introduced in 1949, the Dodge Wayfarer line included a three-passenger Roadster. This early edition has detachable windows. Dodge later changed to roadsters with vent wings and roll-up window glass. (Chrysler Historical)

In the early 1960s, Dodge helped carry the MoPar banner in the sport of drag racing. The 1964 Dodge 880 Ramcharger two-door sedan was available with powerful V-8 engine and factory lightweight aluminum body components.

Ford Highlights

In 1896, the first of 25 pre-production cars were built by Henry Ford in a factory at 58 Bagley Avenue, in Detroit, Michigan. Articles of incorporation were filed in the name of Ford Motor Company, a Michigan corporation, in 1903. The first Ford car was sold that year. In 1907 came the detachable cylinder head, in 1909 the Model T and 1909 the organization of a Delaware corporation. By 1925, Ford was building 10,000 Model Ts per day. The Ford flathead V-8 debuted in 1932 and, by 1938 over five million V-8s were manufactured. The twenty-nine millionth Ford was made in 1941.

Like other makers, World War III found Ford just putting a new style into production when manufacturing was halted in early 1942. The 1946 cars were similar to 1942s, as were all Fords through 1948. The wood-bodied Sportsman convertible was new, however, and became one of the most desirable early postwar models.

February 15, 1949 marked the formation of Ford Division. New that year was a slab-sided car with a slightly revised version of the time-worn L-head V-8. Fordomatic automatic transmission was introduced in 1950 and the Victoria hardtop coupe came along in 1951. The next year was highlighted by the release of an overhead valve V-8 to help Ford keep up with the changing times.

Ford's collector car track record was excellent in the 1950s. The fourth year of the decade brought the Skyliner hardtop with a tinted plexiglas roof insert. It was followed by issuance of the classic two-seat Thunderbird in 1955, the Crown Victoria in 1955-1956 and the Skyliner retractable (a genuine hardtop-convertible) in 1957, the same year the wagon-turned-pickup Ranchero appeared. Headlining 1958 was a new four-passenger Thunderbird. In late 1959 Ford introduced its 1960 models, including the compact-sized Falcon.

Ford's theme in the 1960s was "total performance." Collectible models of this era include Thunderbird Sport Roadsters, Falcon Futura Sprint convertibles, big-block Galaxie 500s, mid-size Fairlane Thunderbolt drag cars, Mustangs, Galaxie 7-Liter rockets, 427 Fairlane GTs and hot Torinos and Talladegas. High-performance lasted until 1972 in the Mustang Mach 1 and Torino Cobra formats and a handful of other variants. 1973 was the final season for a Mustang convertible in the special-interest car era.

This book looks into the finer points of the ever-popular 1939 and 1940 Fords, then swings to the 1950 "shoe box" era. The 1955 Thunderbird is inspected, as well as the Thunderbird-inspired 1958 Fords. The redesigned 1960 Fords and the 1961 Country Squire station wagon are reviewed next, along with the 1961 Thunderbird.

Ford Picks & Prices

(1) 1948 Super Deluxe Sportsman Convertible $48,000
(2) 1949 Custom Deluxe Station Wagon $19,000
(3) 1954 Skyliner Glass-top $13,000
(4) 1959 Skyliner Retractable $24,000
(5) 1964-1/2 Mustang Convertible $23,000
(6) 1965 Thunderbird Convertible $21,000
(7) 1969 Mustang Boss 429 .... $52,000
(8) 1969 Mustang Boss 302 .... $28,000
(9) 1970 Torino Cobra Sportsroof $19,000
(10) 1973 Mustang Convertible $17,000

The Mustang was introduced at the New York World's Fair on April 1, 1964 and paced the Indy 500 that May. Ford built the first Mustangs as 1965 models, but they have distinctions. Collectors sometimes call them 1964-1/2s. (IMSC photo)

In 1973, the last Mustang convertible of the special-interest car era was made. With a standard six-cylinder engine and no options, it sold for $3,102. Only 11,853 of these cars were manufactured.

Hudson Highlights

The Hudson Motor Car Company was organized in 1909. In 1912, the company marketed the first six-cylinder car to sell at a popular price. It proved so popular that the company had to expand its plant to 847,000 square-feet of floor space in 1916. Three years later the four-cylinder Essex was introduced. The Essex was the first low-priced car to offer a closed body and, within a year, the size of the Hudson plant had to be doubled to accommodate the added business it brought. In 1926, Hudson became the first automaker to build bodies on a production basis in its own plant. The Essex-Terraplane line was added in 1932.

Already stylish, Hudsons offered many innovations in the mid- to late-1930s from "Axle-Flex" semi-independent front suspension to "Electric-Hand" vacuum-powered gear shifting. This was also a period in which the firm began to compete in races and speed runs to promote the performance of its products. Hudson had some lean years and styling failures in the mid-1930s, but started to rebound again in 1938, when over 50,000 cars were shipped. New "Symphonic" styling for 1940 pushed shipments to nearly 90,000 that year and got things back on track just before World War II began.

Hudson produced its three millionth car in 1947, the same year it spent $16 million changing over to Monobilt "step-down" body and frame construction. The Pacemaker six was added in 1949. The Hornet came along, in 1950, to light up the race track and the sales race. Hydra-Matic transmission, sourced from General Motors, arrived in 1951 Hudsons and the old Drivemaster and Supermatic transmissions were dropped. The Wasp and the Hollywood hardtop made their bows and the Super Six line was discontinued.

Speaking of racing, Hudsons captured every American Automobile Association (AAA) stock car record in 1952. The compact Jet model was also launched. Power brakes and a new Borg-Warner automatic transmission appeared on some Hudson products in 1953. By 1954, Hudson was beginning to see its stock car racing dominance, as well as its identity, diminish. On May 1 of that year, the Hudson Motor Car Company became a division of American Motors Corporation. With the close of model-year production on October 29, came the end of Hudson assemblies in Detroit, Michigan. From December 28, 1954 through 1957, the Hudson name continued to appear on a series of Nash clones built in Kenosha, Wisconsin. Then the marque vanished forever.

This book features an entry about the 1939 Hudson Country Club models.

Hudson Picks & Prices

(1) 1935 Deluxe Eight Convertible $32,000
(2) 1937 Custom Eight Convertible Brougham $32,000
(3) 1941 Commodore Eight Convertible $31,000
(4) 1949 Commodore Eight Convertible $30,000
(5) 1951 Hornet Convertible $29,000
(6) 1953 Hornet Hollywood Hardtop $13,000
(7) 1954 Jetliner Club Sedan $7,700
(8) 1954 Hornet Brougham Convertible $26,000
(9) 1955 Italia Coupe $28,000
(10) 1957 Hornet Custom Hollywood $14,000

Hudson had lean years and styling failures in the mid-1930s, but "skated" through the end of the decade. The "symphonic" styling of this 1940 Commodore Eight pushed shipments to 90,000 and got things back on track before the war.

Hudson spent $16 million changing over to the Monobilt "step-down" body and frame. The Hornet came along to light up race tracks and put new life in the sales race. This is the 1953 Hudson Hornet coupe.

Imperial Highlights

Imperial was Chrysler's luxury car. It bowed in the 1927-1928 series and included the marque's first true convertible. Longer and lower than its parent, with a larger "Red Head" six, a wide range of factory semi-custom bodies were soon seen on the Imperial chassis. A Custom Imperial line of 1932 included models with longer hoods. Most had semi-custom LeBaron bodies.

In the special-interest era, the Imperial name graced three Airflow lines: Imperial on a 128-inch wheelbase; Custom Imperial on a mid-size wheelbase; and limited-edition CW Custom Imperial on a 146-inch wheelbase. Conventional styling returned in 1938 and a New York Special was added. For 1939, Imperial, New Yorker and Saratoga models were grouped in one series. A Custom Imperial line was offered, too. Beginning in 1940, the New Yorker/Saratoga lines were separated and the largest, most luxurious cars were Crown Imperials.

Crown Imperial was used through 1948. The following year, the Imperial was based on the New Yorker, but had special roof and interior trim items from Derham Body Company. Crown Imperials (basically limousines) were offered, too. Things stayed pat through 1954. Then Imperial became a separate brand from the New Yorker/Imperial Division of Chrysler. Swank new styling for 1955 boosted the series image in the luxury car field, where it captured second place.

In the late 1950s, the Imperial looked more like a Chrysler again. It's image grew confused. For awhile, it was also difficult to figure what the names of models meant. The basic series was called Imperial Custom, the fancier series was Imperial Crown, and the limousines were Crown Imperials. The big cars were really rare. In 1959, for instance, just seven Crown Imperials were built.

Things stayed confused in the early 1960s. There were four series: Imperial Custom, Imperial Crown, LeBaron and Crown Imperial. Despite unique free-standing headlamps and taillamps, a Chrysler look was obvious. Direst mailings in 1962 to "leading doctors" and "eminent attorneys" aided sales very little. At the end of the year, the New Yorker/Imperial Division identity was dropped. Imperials were now made by the Chrysler-Plymouth Division of Chrysler.

Following Lincoln's lead, 1964 to 1968 Imperials looked cleaner and more conservative. Imperial Crown, Imperial LeBaron and Crown Imperial nameplates were offered through 1965, when the low line took the high line's name. For 1967, there were Imperials, Imperial Crowns and Imperial LeBarons. From 1968 to 1970 there were Crown and LeBaron models. Only the Imperial LeBaron survived later. The 1969 to 1973 cars adopted Chrysler's "fuselage" look.

Imperial Picks & Prices

(1) 1937 Custom Airflow Town Limousine $39,000
(2) 1939 Derham Custom Convertible Sedan $60,000
(3) 1951 Convertible $18,000
(4) 1955 Newport Hardtop $18,000
(5) 1959 Crown Convertible $29,000
(6) 1961 Crown Ghia Limousine $18,000
(7) 1963 Crown Convertible $16,000
(8) 1965 Crown Convertible $15,000
(9) 1965 Crown Ghia Limousine $17,000
(10) 1968 Convertible $17,000

The name Crown Imperial was used until 1949, when the all-new postwar Imperial came out. It was based on the New Yorker, but had special roof and interior trim items sourced from Derham Body Company. (Chrysler Historical)

In the late 1950s, it was difficult to figure out the model names. The base series was Imperial Custom, the Imperial Crown was fancier and limousines were Crown Imperials. In 1959, just seven Crown Imperials were built. (Joe Bortz)

King Midget Highlights

In terms of years of continuous production, no strictly postwar American automaker came even close to Midget Motors in terms of longevity. The company built its small cars from 1947 through 1969.

The King Midget was first developed by Claud Dry and Dale Orcuft as a small one-person roadster with the same styling as a midget race car. A six-horsepower, Wisconsin, one-cylinder, air-cooled engine was mounted in the rear and a type of automatic transmission was used (except on very early models). The car had a base price of $270.

By the early 1950s, the Midget became a two-passenger job with a front end that looked like a dodge-'em car and a body that looked like a shrunken Jeep. It gained some $280 in price and 120 pounds of weight (up to 450 pounds). Horsepower was also up to seven and one-half. By 1956 or 1957, another one horsepower was added.

A third type of King Midget styling bowed in 1958. The rear of the one-foot longer body had a hint of tailfins. Wheelbase increased from 72 inches to 76.5 inches. Overall length literally leaped from 102 inches to a "big" 117 inches. Front and rear treads gained four inches (up to 44 inches) and tire size went from 4.00 x 8 to 5.70 x 8. The price increased $300 to $400 depending upon the available equipment ordered, which now included doors, doors with woodgraining, a folding top with side curtains and turn signals. King Midgets now weighed about 675 pounds and had nine and one-quarter horsepower.

The final 1967 to 1969 models were switched to 12-horsepower Kohler engines. They sold for a tad under $1,100 and tipped the scales at 800 pounds. Since life's little pleasures must ultimately end, Midget Motors Corporation decided to fold up its shop in Athens, Ohio after 1969. At that point, the firm had produced some 5,000 automobiles over a period of 24 years. It was no little accomplishment considering how many other postwar automakers disappeared after only a year or two.

King Midget Picks & Prices

(1) 1947 Racer...$3,500
(2) 1950 Racer...$3,200
(3) 1951 Roadster..$3,200
(4) 1957 Roadster..$3,200
(5) 1958 Roadster..$3,500
(6) 1959 Woodgrain Roadster...$3,300
(7) 1966 Roadster..$3,500
(8) 1967 Roadster..$3,500
(9) 1968 Roadster..$3,500
(10) 1969 Roadster..$3,500

A third type of King Midget styling bowed in 1958. The rear of the one-foot longer body had a hint of tailfins. Wheelbase increased from 72 inches to 76.5 inches. Overall length literally leaped from 102 inches to a "big" 117 inches.

Lincoln Highlights

Lincoln Motor Car Company was incorporated in the state of Michigan in 1917. It was originally formed to manufacture Liberty engines. In 1920, the firm was reorganized to build cars and incorporated in the state of Delaware. Ford purchased Lincoln two years later.

Lincoln-Zephyr production began in 1935 and the Lincoln Continental was introduced in 1939. It wasn't until October 1945 that the name Lincoln-Mercury Division evolved for this branch of Ford Motor Company. In 1948, the Lincoln V-12 was replaced by a V-8, the Continental was temporarily discontinued, the Cosmopolitan was introduced and Lincoln established an all-time record with production of 43,938 cars. Lincoln consolidated its 1949 production in plants in Detroit and Los Angeles. Also that year, Hydra-Matic automatic transmission, sourced from General Motors, was offered.

Lincoln began construction of a plant in Wayne, Michigan in 1950. The following year, its 121-inch wheelbase series was dropped. In October 1952, assemblies at the Detroit factory were discontinued. Nearly a half-million Lincolns had been built there. The same year, Ford established a Special Products Operations branch with William C. Ford as manager. Production of 1953 models with a new 205-horsepower four-barrel V-8 then began at the new Wayne plant. The Ford Motor Company board of directors also approved a program to launch a new marque named Continental. By November 20, 1953 the 250,000th postwar Lincoln rolled off an assembly line at Wayne. Three days later, Lincolns took the top four spots in the Pan-American Road Race in Mexico.

Ground was broken for a new plant, in Ecorse Township, Michigan in 1954. The new Continental would be built there. Also completed and delivered to the Continental Division that year was the first engineering prototype of the Continental Mark II. On April 15, 1955, Lincoln became a separate division of Ford, as did Mercury. A separate Continental Division was also formed to build and market the new Continental Mark II. Production versions of the new luxury car left assembly facilities in June 1955 and were introduced to the public on October 21.

The 1958 to 1960 Lincolns were the biggest American postwar cars. Distinctions between Lincolns and Continentals was basically in trim and body style offerings. All lines used the same engine and wheelbase. These cars were originally designated Mark IIIs, IVs and Vs. This was officially forgotten when Lincoln reissued the same designations later. Apparently, they wanted to forget the huge, wide, flat cars. The 1961 Continental was a refreshing return to clean lines and tasteful use of chrome. An exciting new model was a four-door convertible. Though enlarged a bit, the clean, slab-sided look lasted through 1967.

In the spring of 1968, the 1969 Continental Mark III was introduced. It looked like a Mark II that had gone to Hollywood. This car proved to be very popular and successful and its design influenced Lincoln and Continental styling through the 1970s. The Lincolns looked at closely inside are 1938, 1951, 1956 and 1969 models.

Lincoln Picks & Prices

(1) 1936 K Brunn Convertible Victoria $90,000
(2) 1940 Continental Cabriolet $64,000
(3) 1947 Continental Cabriolet $45,000
(4) 1951 Cosmopolitan Convertible $26,000
(5) 1953 Capri Convertible $26,000
(6) 1957 Continental Mark II $29,000
(7) 1961 Continental Phaeton $14,000
(8) 1963 Executive Limousine $12,000
(9) 1967 Continental Phaeton $14,000
(10) 1972 Continental Mark IV $11,000

By November 20, 1953 the 250,000th postwar Lincoln rolled off an assembly line at the Wayne, Michigan factory. Three days later, Lincolns took the top four spots in the Pan-American Road Race. This is a 1953 Cosmopolitan hardtop.

The 1958 Continental was a big American postwar car. Distinctions between Lincolns and Continentals diminished and all lines used the same engine. These cars were originally designated Mark IIIs, but this was officially forgotten.

Mercury Highlights

The first Mercury was introduced in October 1938 to fill the gap between the Ford Deluxe and the Lincoln-Zephyr. It was originally conceived of as a "fancy Ford" and did well through February 10, 1942, when civilian auto production stopped for World War II. On November 1, 1945, assemblies of prewar-style postwar Mercurys began. The Lincoln-Mercury Division of Ford was also formed that year. Later Mercurys would have more of a "baby Lincoln" image.

On April 29, 1948 the all-new postwar Mercurys ... best known now as "James Dean" models because the legendary actor drove one in a film ... made their bow and sparked a more than seven-fold increase in sales. This inspired a need for new assembly plants in Los Angeles, California, St. Louis, Missouri, and Metuchen, New Jersey in 1949 and Wayne, Michigan in 1950. In March 1952, Mercury took over the Detroit Lincoln factory until the Wayne Plant got rolling.

Mercury introduced power brakes on April 10, 1953 and power steering came one month later. On August 17 of the year, the 1.5 millionth postwar Mercury was built. The Metuchen Plant was expanded in 1954, the year that Mercury launched the glass-top Sun Valley hardtop. In 1955, Mercury became a separate division of Ford Motor Company and roared to production and sales tempos unprecedented in its history.

The Ford-based mid-century "Mercs" looked distinctive. In 1957, the marque got its own body, which wasn't shared with other FoMoCo nameplates. Unusual shapes and gobs of chrome characterized body designs through 1960, when the compact Comet sent Mercury heading in a smaller new direction. Next came the mid-size Meteor and the re-introduction of the "Breezeway" rear window (first released in 1957). The small Mercurys were especially popular with young fans of Ford's "Total Performance" push, while full-size fastbacks, like the Marauder, appealed to more mature speed merchants.

The Lincoln Continental tradition was apparent in full-size Mercurys for 1965 when Lee Iacocca was charged with creating a new sales-building program. For 1967, the big news was the Cougar, which was like a Mustang dressed in a tuxedo. In 1968, the Montego joined the Comet as a fancier mid-size model. It came in a hot Cyclone GT fastback, too. The early 1970s saw an enlargement and restyling of the Cougar, a T-bird-like Monterey and the introduction of the sub-compact Bobcat. The final special-interest Mercury ragtops, both Cougars, were seen in 1973.

This book examines the 1949 Mercury "woodie" station wagon and the 1970 Marauder X-100 high-performance model. Here's some other collectible models to look into.

Mercury Picks & Prices

(1) 1939 99A Convertible	$27,000
(2) 1947 Sportsman Convertible	$46,000
(3) 1949 OCM Coupe	$10,000
(4) 1954 Sun Valley Coupe	$17,000
(5) 1957 Turnpike Cruiser Convertible	$29,000
(6) 1963 Comet S-22 Convertible	$12,000
(7) 1963 Monterey S-55 Convertible	$13,000
(8) 1966 Comet Cyclone GT Convertible	$13,000
(9) 1968 Cougar GTE Hardtop	$13,200
(10) 1969 Cougar Eliminator Hardtop	$14,500

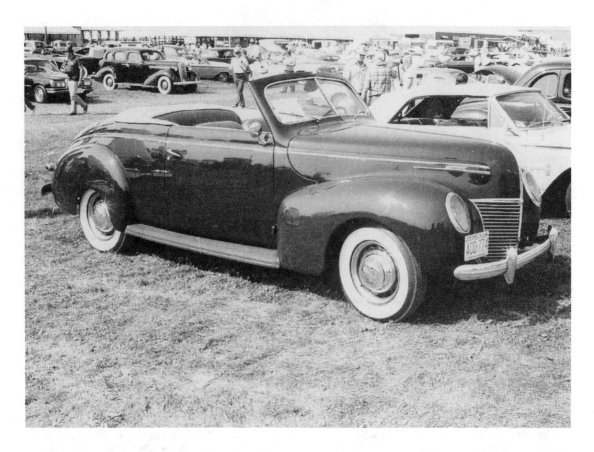

The 1939 Mercury was introduced in October 1938 to fill the gap between the Ford Deluxe and the Lincoln-Zephyr. It was originally conceived of as a "fancy Ford" and did well until civilian auto production stopped for World War II.

In 1957, the marque got its own body, which wasn't shared with other Ford Motor Company nameplates. Unusual shapes and gobs of chrome characterized the new Mercury's styling. (Ford Archives)

1942 Plymouth Special Deluxe convertible

1946 Ford Super Deluxe V-8 *station wagon*

1951 Hudson Hornet Convertible Brougham

1955 Chevrolet Bel Air Sport Coupe *two-door hardtop*

1957 Ford Fairlane 500 Sunliner convertible

1964 Ford Galaxie 500XL Club Victoria *two-door hardtop*

1968 AMX *fastback coupe*

1971 Oldsmobile 4-4-2 *convertible*

Oldsmobile Highlights

The first Oldsmobile was produced way back in 1897 and Olds Motor Vehicle Company was formed that same year. In 1899, the company opened the first plant exclusively for the production of automobiles. In 1905, all production was moved to Lansing, Michigan. Three years later, Oldsmobile joined General Motors Corporation. A new assembly plant was opened in 1910. A V-8 model was introduced in 1916. In 1917, the factory was converted to war production. A new engine plant was constructed in 1918.

Oldsmobile settled down to building only sixes until 1929. It introduced the use of chrome plated parts in 1925. The Viking companion car brought a V-8 back for two years, but lasted only until 1930. It was sixes only in 1931 and straight sixes and eights thereafter until World War II ended. The 1,000,000 Oldsmobile was built in 1935, the same year a $6 million expansion was begun. An important first for the division in 1939 was Hydra-Matic drive. During the war, Oldsmobile made aircraft parts, 75- and 105-millimeter shells, forgings and canon.

The Rocket V-8 engine was introduced in late 1948. Sales more than doubled by 1949. The 3,000.000 Oldsmobile was made in February 1950. By 1951, the company was back in the business of manufacturing defense goods for the Korean War effort. Construction of a plant for making rotating parts for J-65 jet engines began in July 1951. For 1952, new Oldsmobile features were an Autronic-Eye automatic headlamp dimmer and power steering. Amazingly, the 1,000,000th Rocket engine was built. Air conditioning was also introduced. A fire at the Hydra-Matic transmission plant impacted sales in 1953, when a special Fiesta convertible was released. The following year, Oldsmobile became the third best selling General Motors marque. A huge expansion program, the largest in company history, got underway in 1955. It increased production capacity by 50 percent. The 1956 models boasted a new automatic transmission, flairaway fenders, projectile taillights and an airfoil-type grille.

Oldsmobile entered the great American horsepower race in 1957 with its triple-carburetor J-2 package. It was also available in the chrome-laden 1958 Oldsmobile. The early 1960s brought the F-85 compact (available in sporty Cutlass dress) and the Starfire personal/luxury car. A creative chief engineer named John Beltz tinkered with technical goodies from aluminum-block V-8s to turbocharging for enhanced performance, but the back-to-basics 4-4-2 of 1964 took the "factory hot rod" approach of stuffing a big-block motor in a now mid-sized Cutlass body.

Innovation was not discarded, though, as evidenced by the 1966 front-wheel-drive Toronado. An exciting high-performance package for 1968 Oldsmobile buyers was the semi-aftermarket Hurst/Olds. By the 1970s, the Cutlass grew into a powerful force in the battle to attract youth market buyers. With new Colonnade hardtop styling, it became the darling of the post-performance era set seeking opera windows and velour upholstery in place of four-speed gearboxes and 400-cube V-8s.

This book looks at the 1942 Oldsmobile that helped Americans on the homefront get around, the glitzy 1958 model and the 1973 Ninety-Eight. Other neat models are listed below.

Oldsmobile Picks & Prices

(1) 1941 Series 68 Station Wagon $28,000
(2) 1949 Dynamic 88 Convertible $26,000
(3) 1953 Fiesta Convertible $45,000
(4) 1957 Super 88 J-2 Convertible $34,000
(5) 1962 Jetfire Hardtop $9,000
(6) 1963 Starfire Convertible $24,000
(7) 1964 Cutlass 4-4-2 Convertible $11,000
(8) 1966 Toronado Custom Coupe $10,500
(9) 1968 Hurst/Olds Coupe $16,000
(10) 1970 Cutlass Rallye 350 Coupe $13,500

FRONT VIEW OF 1951 OLDSMOBILE SERIES "98" HOLIDAY

By 1951, Oldsmobile was back into defense goods manufacturing for the Korean War effort. Construction of a plant for making rotating parts for J-65 jet engines also began in July 1951. This is the "98" Holiday coupe. (Oldsmobile)

By the 1970s, the Cutlass grew popular with young buyers. This 1973 Cutlass Salon has Colonnade hardtop styling. It was the darling of post-performance era buyers seeking opera windows and velour upholstery. (Oldsmobile)

Packard Highlights

The first Packard dates to 1899 and Packard Motor Car Company formed in 1902. A famous Packard was the 12-cylinder Twin Six of 1915. In 1929, the company produced the first radial aircraft engine. A new V-12 appeared in some of the classic Packards of 1932 and later vintage. Also available were Standard Eight, Custom Eight, and Super Eight series.

As the special-interest era started in 1935, Packard issued the medium-priced 120 series and enjoyed initial success with it. A six-cylinder line was added in 1937. It helped make Packard ownership a reality for a new class of buyers, but may have tarnished the company's high-end image a bit. Introduced first in 1938 was the sporty Packard-Darrin convertible with cut-down doors.

Packard resumed passenger car production in 1946. New and larger cars were introduced in 1948 as 1949 models. Packard also celebrated its 50th anniversary in 1949 and Ultramatic transmission was introduced that year. The new Packards were dubbed "pregnant Packards" due to their smooth, bulbous body lines. Traditional Packard buyers did not embrace them very warmly. Sales declined by about 40,000 units from 1948.

The early 1950s saw a series of product and facility expansions. Sales continued to drop, by another 6,000 cars, to a total of 100,312. New management took over in 1952, when total model-year production was 62,922 cars. Late in 1952, the low-priced Clipper was reintroduced. In 1953, the Detroit automaker sold 61,242 Packards and 30,035 Clippers and resumed building custom limousines and sedans.

Things looked rosy, but not for long. Model-year output for 1954 was 21,208 Clippers and 10,106 Packards. In October, Packard merged with the Studebaker Corporation of South Bend, Indiana. The 1955 to 1956 cars were true Packards with a new torsion-bar front suspension system, push-button Ultramatic transmission, and overhead valve V-8 engines producing up to 310 horsepower in Caribbeans. Advanced safety features of these cars included interlocking safety door latches, seat belts and foam rubber instrument panel pads. For 1957 and 1958, the Packard name was switched to upscale Studebakers with special power trains. Some collectors say the last REAL Packards were built in 1956; others argue that the models of the last two years are also special-interest cars. This book examines 1953, 1955 and 1957 Packards.

Packard Picks & Prices

(1) 1936 Super Eight Convertible $136,000
(2) 1940 Custom Super Eight Darrin Convertible $136,000
(3) 1947 (21st Series) Clipper Super Eight Limousine...$29,000
(4) 1948 (22nd Series) Custom Eight Convertible..$38,000
(5) 1950 (23rd Series) Custom Eight Convertible $35,000
(6) 1951 (24th Series) Convertible..$21,000
(7) 1953 (26th Series) Caribbean Convertible..$29,000
(8) 1955 (55th Series) 400 Hardtop Coupe...$21,000
(9) 1956 (56th Series) Caribbean Convertible..$34,000
(10) 1958 (58th Series) Hawk Hardtop $24,000

As the special-interest era began, Packard issued the medium-priced 120 series. However, it continued to offer upscale cars like this 1936 Super Eight Model 1404 Convertible Coupe. (Applegate & Applegate photo)

New management took over at Packard Motor Car Company in 1952, when total model-year production was 62,922 cars. Late in the calendar-year, the low-priced Clipper was reintroduced as a 1953 model. This is a Custom Clipper Deluxe.

Plymouth Highlights

The first Plymouth was built June 14, 1928. Four-wheel hydraulic brakes and a 45-horsepower four-cylinder engine were highlights of the inexpensive new Chrysler Corporation marque. In 1930, Chrysler, DeSoto and Dodge dealers began selling the new line, which led to a doubling of sales and a 25 percent factory expansion. By 1931, Plymouth held the industry's number three rank in production and sales. Introduced that season was the "floating power" engine mounting system.

Plymouth built its one millionth car in 1934. A power-operated convertible top was a 1939 innovation. By 1941, total production for the marque crested four million cars. Like other automakers, Plymouth contributed to World War II by producing military orders. Early postwar cars were warmed-over 1942s. New for 1949 was the Suburban, a new type of all-steel station wagon.

By 1951, Plymouth built its seven millionth car. A new Belvedere hardtop was also seen. The 1953 models set an all-time output record of 662,515 for the calendar-year; 636,000 for the model-year. Completely restyled Plymouths with an optional overhead valve V-8 were seen in 1955. The first fully-automatic Chrysler transmission, called PowerFlite, was available in 1953 models.

1956 brought tailfins to Plymouth, plus push-button gear shifting. The hot-trotting Fury was released in 1956 with special gold and white trim and a Canadian V-8 tricked-out with special performance hardware. The 1957-1959 models were clean-styled, but carried wild fins, quad headlights (only where legal in 1957) and, sometimes, spare tire embossments on the trunk.

Plymouth experienced sales declines in the 1960s. Highlights of that decade included release of the "slant six" engine; "factory lightweight" competition packages for drag racing and stock car racing; "Street Hemi" engines, also for racing; and new high-performance models, such as the Satellite, the GTX, the Road Runner and the Super Bird. The early 1970s saw most attention going to muscle car variants of the mid-size Satellite and the sports-compact 'Cuda. With the dropping of convertibles and hemi engines after 1971, Plymouth's specialty models seemed a lot less special and far less interesting. In this book are stories about 1950, 1956 and 1969 Plymouths.

Plymouth Picks & Prices

(1) 1939 Deluxe Convertible Sedan.. $22,000
(2) 1948 Special Deluxe Station Wagon $13,000
(3) 1958 Fury Convertible $18,000
(4) 1960 Fury Convertible $12,000
(5) 1961 Fury Convertible $10,500
(6) 1964 Barracuda $10,500
(7) 1967 GTX Convertible... $17,000
(8) 1970 Road Runner Super Bird $55,000
(9) 1971 Hemi 'Cuda Convertible $34,000
(10) 1973 Sebring Satellite Coupe $11,000

The high-performance 1967 Plymouth GTX featured a "Pit-Stop" fuel filler cap, Red Streak tires, dual hood scoops and sport stripes, heavy-duty transmission and suspension and a standard 440-cubic-inch V-8. (Chrysler Historical)

Pontiac Highlights

The Pontiac Motor Division of General Motors evolved from the Pontiac Buggy Company, which was formed in 1893 in Pontiac, Michigan. It became the Oakland Motor Car Company in 1907 (Oakland being the county name) and part of General Motors in 1908. The first Pontiac car was an inexpensive six-cylinder coupe created for marketing by Oakland automobile dealers beginning in 1926. It proved so popular that, by 1932, Oakland was entirely eclipsed by Pontiac.

At the start of the special-interest era, Pontiac added chrome "Silver Streaks" to its hood. These were a company trademark through 1956, a year after Pontiac released an overhead valve V-8 and set off on a high-performance binge. In 1957, a young new general manager named "Bunkie" Knutson took over Pontiac and aimed it at the youth market. He sired a limited-production fuel-injected convertible called the Bonneville, a hardtop wagon called the Custom Safari and a popular new line of "Wide-Track" Pontiacs for 1959.

During the early 1960s, Pontiac became a force in both stock car competition and drag racing. The wins on Sunday led to car sales on Monday and Pontiac climbed into third rank in auto industry sales for the full decade. A small Tempest compact car was unveiled in 1961 and grew into a mid-size high-performance model called the GTO. This car set off the muscle car stampede, which was joined by the full-sized Catalina 2+2 and Bonneville 428 models. Another hot product was the Grand Prix, a sports/luxury hardtop (and convertible in 1967-only) which went through several successful restylings and still survives as a sales leader today.

1972 brought an end to the high-performance years and signaled a slight decline in Pontiac's fortune. This period was marked by the near-death of the Firebird (before it sparked a sales revival) and the end of Pontiac convertible production in 1975. There were attempts to make the GTO a performance-image version of a new Ventura compact, before sending the nameplate to the "muscle car hall of fame" after 1974. Pontiac rode out the 1970s floundering, before revitalizing its identity and purpose at the close of the decade. This book researches the 1937, 1941, 1963 and 1969 Pontiacs. Here's a few more to keep your eyes on.

Picks & Prices

(1) 1937 Deluxe Eight Convertible Sedan..$28,000
(2) 1948 Deluxe Torpedo Eight Convertible $22,000
(3) 1953 Chieftain Deluxe Eight Convertible..$22,000
(4) 1957 Star Chief Bonneville Custom Convertible $54,000
(5) 1958 Bonneville Convertible..$36,000
(6) 1964 LeMans GTO Convertible..$18,000
(7) 1966 Catalina 2+2 convertible $13,000
(8) 1967 Grand Prix Convertible $14,000
(9) 1970 GTO Judge Convertible $30,000
(10) 1973 Trans Am SD-455 Coupe $15,000

Pontiac offered a convertible sedan for just two years, starting with the 1937 model. This body style was available on both the Deluxe Six and Deluxe Eight chassis. (Pontiac photo)

The straight eight engine introduced in 1933 was in its next-to-last appearance during 1953, when Pontiac modified its traditional early postwar styling with small, rounded "fins" on the rear fenders. (Pontiac photo)

Shelby Highlights

In the early 1960s, race car driver Carroll Shelby got the idea that dropping a Ford V-8 into the aluminum-bodied British AC Ace sports car would be fun. By 1962, the Shelby Cobra roadster was on the market at $5,995. With 260-cubic-inch, 260-horsepower V-8 and 2,020-pound chassis weight, the car was brutally fast. That year, about 75 Cobras were made by Shelby-American, Incorporated of Venice, California.

For 1963, the only switch was a change to the 289-cubic-inch Ford V-8. It was good for 271 horsepower. The 289 Cobra was offered through 1965 and a total of some 580 were made.

Since the small-block-in-small-car formula had worked so well, Shelby got the idea to stuff the 427 cubic inch Ford big-block V-8 into the 1965 Cobra. This put 425 horsepower on tap and created a street racer unsuitable for the feint of heart. The 427 Cobras remained available through 1967 and some 356 were sold.

While continuing to make 427 Cobras, Shelby-American Automobiles also began modifying Ford Mustangs for high-performance. These Shelby Mustangs had special scoops, stripes and heavy-duty equipment and cost over $4,500, which was a lot at the time. A 306 horsepower version of the 289-cubic-inch V-8 was the power plant. The GT 350 version was offered as a fastback only in 1965. In 1966, there were GT 350 and GT 350H (for Hertz-Rent-A-Car) fastbacks and a GT 350 convertible. The 936 cars built for Hertz were often rented by enthusiasts for weekend excursions to the racetrack. A GT 500 model with a 428-cubic-inch V-8 was introduced as a fastback-only in 1967. A GT 500KR (for King of the Road) version was new and exclusive for 1968. The final year was 1970, when partially-built 1969 models were completed and sold.

Shelby Picks & Prices

(1) 1962 Cobra "260" Roadster	$190,000
(2) 1963 Cobra "289" Roadster	$200,000
(3) 1965 Cobra "427" Roadster	$320,000
(4) 1965 GT 350 Fastback	$43,000
(5) 1966 GT 350H Fastback	$39,000
(6) 1967 GT 350 Fastback	$30,000
(7) 1968 GT 350 Convertible	$50,000
(8) 1968 GT 500 Convertible	$66,000
(9) 1969 GT 500KR Fastback	$33,000
(10) 1970 GT 500 Convertible	$68,000

Among the rarest of the rare are the handful of Shelby-Daytonas made for GT sports car racing. These powerful coupes looked mean and nasty and had the horsepower necessary to live up to their eye appeal. (Carroll Shelby)

Studebaker Highlights

It was way back in 1852 that H. and C. Studebaker founded their blacksmith and wagon making shop. Six years later, it was recapitalized as the C. & J. M. Studebaker Company. Studebaker Brothers Manufacturing Company was incorporated 10 years later. By 1902, the company was making electric runabouts and trucks. The Studebaker Corporation was established in South Bend, Indiana in 1911. It operated plants there and in three Michigan cities: Port Huron, Pontiac and Detroit. By 1926, when some other auto companies were just getting going, Studebaker had built its one millionth motor vehicle.

The Great Depression hurt the company. By 1933, it was in receivership. A reorganization of 1935 saved the day for the next 30 years. During 1939, Studebaker introduced its low-priced Champion and built its two millionth car. Then, World War II came along and the emphasis switched to making trucks and implements of war.

Studebaker was the first automaker to release an all-new postwar car. It bowed in 1946. This up-to-date car was a fashion sensation and sales skyrocketed after its appearance. In 1948, Canadian operations resumed. More capacity was needed to produce sufficient numbers of cars to meet the strong demand. The one millionth postwar Studebaker rolled off an assembly line in 1950, the same year that a V-8 engine and automatic transmission appeared. Naturally, a Studebaker hardtop came along quickly in 1951.

Celebrating its centennial in 1952, Studebaker converted to the manufacture of the first American-built production car with imported sports car styling. This Raymond Loewy-designed Starlight Coupe bowed to rave reviews in 1953. The Museum of Modern Art contemporarily recognized its contribution to automotive styling. In October 1954, Studebaker and Packard joined forces through a merger.

A new Hawk series began in 1956, as the passenger cars got a new body. The first in the series were powered by a big Packard-built V-8. The 1957 Golden Hawk had a supercharged 289-cubic-inch V-8. 1959 brought the Lark to market. It was a compact, yet roomy, box of a car that set the stage for Studebaker's finale. However, the company had one blast of creativity left and expended it on the design and development of the fiberglass-bodied Avanti, which continues to be built today. This book features 1935 and 1955 Studebakers. Here's some more.

Studebaker Picks & Prices

(1) 1941 President Skyway Land Cruiser $16,300
(2) 1948 Commander Regal Convertible $19,000
(3) 1953 Commander Regal Starliner Coupe $12,500
(4) 1956 Golden Hawk Hardtop $18,000
(5) 1958 Golden Hawk Sport Hardtop $16,000
(6) 1962 Grand Turismo Hawk Hardtop $13,000
(7) 1963 Lark Daytona Convertible $10,500
(8) 1964 Avanti Coupe $19,000
(9) 1965 Cruiser Sedan $5,700
(10) 1966 Avanti II $20,000

Studebaker introduced its low-priced Champion and built its two millionth car just before World War II. One of the prettiest was the 1941 Studebaker President Skyway Cruising sedan. (Applegate & Applegate photo)

Studebaker offered the first American production car with imported sports car styling. The Raymond Loewy-styled Starliner Coupe bowed to rave reviews in 1953. The Museum of Modern Art recognized its contribution to car design.

Tucker Highlights

In the mid-1940s, Preston Tucker began assembling a staff to help him build his dream car. He was well-known in automotive circles as a good salesman and he was determined to make a revolutionary contribution to the auto industry. Alex Tremulis, a talented designer who had worked on Cords and Duesenbergs had a metal prototype ready to go in 100 days. The car was sensationally styled, lower than any car then in production and featured a much wider track. The mechanics of the new car. It was really intriguing.

The engine was a 334.1-cubic-inch flat six built primarily of aluminum and mounted in the rear. It developed 150 horsepower, making the Tucker one of the fastest cars on the American road. Safety was an obsession with Preston Tucker. The body structure was extremely solid and the frame was stronger than that of any other contemporary American car. Crash padding was used on interior surfaces and the windshield was designed to pop out on impact.

From a standpoint of specifications, the Tucker was a real contender in 1948. It was the fastest sedan on the market, with 0 to 60 times under 11 seconds and a top speed over 100 miles per hour. It regularly delivered over 20 miles per gallon. This is borne out by the owners of the 51 cars that were eventually placed in service. Interior accommodations were the equal of the Cadillac. Doors that opened into the roof area eased entry and exit into the unusually low machine. Three headlamps lit the road ahead and the center-mounted one turned with the front wheels. Tucker was full of good ideas, unconventional, but good.

With articles in almost every major magazine and newspaper, the market for Tucker's new car seemed assured. There was tremendous public interest and a large number of orders were placed. Purchased from Chrysler Corporation, a plant in Chicago was all ready to begin volume production when the government pulled the plug. The Securities and Exchange Commission chose the beginning of the Tucker experiment as a good time to scrutinize the financial backbone of the fledgling automaker.

Accusations that fraudulent stock deals were being made were wildly flung about and it didn't take long for a confidence crisis to develop. Investors quickly became reluctant to put their money into a company that was considered suspect by the American government. The case went to trial in 1950 and Tucker was acquitted of all charges filed by the over-zealous government agency.

Of the 51 cars that were produced, a great majority of them are still in service or in museums. The car proved its worth and each surviving example retains that arrogant, independent flair that Preston Tucker admired so.

Picks & Prices

There is only one Tucker model. See the review in the next section of this book.

One of the cars used in the movie "Tucker: The Man and His Dream" on exhibit at the "Wild Kingdom" in Silver Springs, Florida. The film was produced by George Lucas and Francis Ford Coppola about Preston Tucker's life.

Tuckers were brought to a courthouse in Chicago, during Preston Tucker's trial, to show that Tuckers worked and that charges of stock fraud were unwarranted. (Wally Wray photo)

Special-Interest Car Review

1968-1969 AMX: built for speed

By Gerald Perschbacher

It came roaring down the track on February 24, 1968, complete with a choice of a 290-cubic-inch, 225-horsepower V-8; an optional 343-cubic-inch, 280-horsepower V-8; or a 390-cubic-inch, 315-horsepower V-8. It set 106 world speed records. It was the 1968 AMX by American Motors Corporation.

This little two-seat car became an important element in AMC's revised approach to styling and marketing. The Kenosha, Wisconsin automaker's focus was switched to promoting speed and handling through both exterior design and performance engineering.

Constructed on a short 97-inch wheelbase, as was the Javelin, the AMX was designated series 30 for two-passengers. It came only as a two-seat fastback with a factory price of $3,245. It tipped the scales at 3,097 pounds. Partly because it was a midyear entry, only 6,725 units were made.

The 1969 version had a few changes. Among them were a 140 miles per hour speedometer and a new tachometer with a larger face. Later in the year, a hooded dash panel was available. New conveniences included a package tray located between the seats and an assist handle for the passenger.

Standard AMX equipment included the four-speed manual transmission mounted through the floor and the four-barrel carburetor 225-horsepower engine. The price for the 1969 version was $3,297, or $52 higher than the original. Also up slightly over year one was the production total, which peaked at 8,293 units.

The two-seat AMX was a showroom traffic-getter that drew buyers into AMC dealerships across the country. It was also a real traffic beater at any "stoplight grand prix." The AMX for 1968 and 1969 was a hard act to follow. However, most cars had no other choice, but to follow it.

SPECIFICATIONS

Year	1968
Make	American Motors Corporation
Model	AMX
Body style	two-door hardtop
Base price	$3,245
Engine	OHV V-8
Bore x stroke	3.75 x 3.50 inches
CID	290
Compression ratio	9.0:1
Carburetor	Carter AFB 4V
H.P.	225 @ 4700 rpm
Wheelbase	97 inches
Overall length	177.2 inches
Weight	3,097 pounds
Tires	E70-14
OCPG Value	$14,000

SPECIFICATIONS

Year	1969
Make	American Motors Corporation
Model	AMX
Body style	two-door hardtop
Base price	$3,297
Engine	OHV V-8
Bore x stroke	3.75 x 3.50 inches
CID	290
Compression ratio	9.0:1
Carburetor	Carter AFB 4V
H.P.	225 @ 4700 rpm
Wheelbase	97 inches
Overall length	177.2 inches
Weight	3,097 pounds
Tires	E70-14
OCPG Value	$14,000

American Motor's first AMX, introduced in midyear 1968, set 106 world speed records. Javelin Racing Team, Incorporated also raced Javelins in Sports Car Club of America Trans Am competition. (AMC photo)

The early AMX race cars are being collected and restored for vintage racing today. The two-seat AMX represented the first important element in AMC's revised approach to styling and marketing. (Vince Ruffalo Collection)

Constructed on a short 97-inch wheelbase, the AMX was designated Series 30 for two-passengers. It was a formidable contender in drag racing too. (Chris Halla photo)

The 1969 AMX had a few changes like a 140 miles per hour speedometer, a new tachometer with a larger face and, later, a hooded dash panel.

1936 Buick 8 "as positive as steel on steel"

By John A. Gunnell

"The notable thing about the ease with which the new Buicks perform every duty is the firmness of that ease," said a 1936 advertisement for one of Flint, Michigan's best-known products. "Though maneuverable under the lightest touch, so that any woman can drive one all day long without fatigue, control is as positive as steel on steel."

Buick historians are unanimous in regarding the 1936 models as cars that marked the start of a renaissance for the automaker. There were new features across the board, including Turret-Top bodies, hydraulic brakes and dramatic new styling with sharply slanted V-type windshields.

Buick's radiator grille had a new "wedge" shape and there were stylishly streamlined twin taillamps and bullet-shaped headlamp buckets. Mechanical improvements included suspension refinements, new alloy pistons and a more effective cooling system.

In addition to numerical designations, each Buick series had its own name. There was the Special for folks just entering the lower regions of the mid-priced market, the Century for the go-fast crowd, the Roadmaster for the conservative businessman and the Limited for the limousine-level customer.

"When better cars are built, Buick will build them," the company bragged in advertisements that highlighted the features and benefits of Buick ownership. The latter included a valve-in-head straight eight engine; anolite pistons; sealed chassis; all-steel body construction; "Tiptoe" hydraulic brakes; "Knee-Action" comfort and safety; torque tube drive; automatic starting, spark and heat control; built-in luggage compartment; and front end ride stabilizer.

There were only two models available in the Roadmaster series. One was a Convertible Phaeton. The other was a sharp-looking four-door Touring Sedan. Both rode on the same wheelbase and used the larger of two eight-cylinder inline engines that Buick manufactured. With five main bearings, the engine featured overhead valves and mechanical valve lifters. Generating 238 pounds-feet of torque, it could motivate the Roadmaster to highway speeds and keep it there all day long.

Hooked to the motor was a three-speed manual transmission with floor shift and a single dry plate clutch. The semi-floating rear axle had a 4.22:1 overall gear ratio. Sixteen-inch-diameter pressed steel wheels and hydraulic brakes all around were other attractions.

The Roadmaster Touring Sedan Model 81 was a popular purchase in the upper-medium price bracket, generating 14,985 assemblies for the model year. That compared to 1,064 for the Model 80C Convertible Phaeton. In addition to all those standard goodies, buyers could pay a bit extra to add one of two heaters (Master or Deluxe), dual side mounted spare tires, fog lights, whitewall tires, a grille guard, an electric watch (clock), a Buick Master five-tube radio or a Ranger six-tube radio, and bright wheel trim rings.

SPECIFICATIONS

Year	1936
Make	Buick
Model	Roadmaster
Body style	four-door Touring Sedan
Base price	$1,255
Engine	OHV; Inline 8
Bore x stroke	3-7/16 x 4-5/16 inches
CID	320.2
Compression ratio	5.9:1
Carburetor	Stromberg 2V
H.P.	120 @ 3200 rpm
Wheelbase	131 inches
Overall length	210.25 inches
Weight	4,098 pounds
Tires	16 x 7.00
OCPG Value	$16,000

The notable thing about the ease with which the new Buicks perform every duty is the firmness of that ease—though maneuverable under the lightest touch, so that any woman can drive one all day long without fatigue, control is positive as steel on steel.

Buick 8
A GENERAL MOTORS PRODUCT

NO OTHER CAR IN THE WORLD HAS ALL THESE FEATURES: *Valve-in-Head Straight-Eight Engine . . . Anolite Pistons . . . Sealed Chassis . . . Luxurious "Turret Top" Body by Fisher with Fisher No Draft Ventilation . . . Tiptoe Hydraulic Brakes . . . Knee-Action Comfort and Safety . . . Torque-Tube Drive . . . Automatic Starting, Spark and Heat Control . . . Built-in Luggage Compartments . . . Front-End Ride Stabilizer*

$765 to $1945 are list prices of the new Buicks at Flint, Mich. Standard and special accessories groups at extra cost

Buick historians are unanimous in regarding the 1936 models as cars that marked the start of a renaissance for the automaker. This contemporary advertisement promotes control and handling.

The 1936 Roadmaster was perfect for the pinstripe-suited businessman who was moving up in the world and had clients to impress. This four-door sedan belongs to Orville Melchiar of New Jersey.

The 1936 Buick Special Touring Sedan was a popular purchase in the lower-medium price bracket. This example in the Blackhawk Auto Collection has 1,200 original miles on its odometer.

So hushed is the oil-cushioned action of the great Buick valve-in-head engine that even in full flight this marvelous car seems "ghosting" along — its silky mobility matched only by the ease and certainty of its fingertip control.

Buick 8

A GENERAL MOTORS PRODUCT

$765 to $1945 are list prices of the new Buicks at Flint, Mich. Standard and special accessories groups at no extra cost

NO OTHER CAR IN THE WORLD HAS ALL THESE FEATURES

Valve-in-Head Straight-Eight Engine . . . Anolite Pistons . . . Sealed Chassis . . . Luxurious "Turret Top" Body by Fisher with Fisher No Draft Ventilation . . . Tiptoe Hydraulic Brakes . . . Knee-Action Comfort and Safety . . . Torque-Tube Drive . . . Automatic Starting, Spark and Heat Control . . . Built-in Luggage Compartments . . . Front-End Ride Stabilizer

There were only two models available in the Roadmaster series. One was a Convertible Phaeton. The other was the sharp-looking four-door Touring Sedan. This advertisement highlights Buick's oil-cushioned engine valves.

Rear view shows the smooth, long trunk and neat taillamps. Inside was an engine-turned dashboard, upgraded interior (with more color choices) and white plastic decorative trim in more places than ever before.

1940 Buick: new features and modern styling

By John A. Gunnell

Buick had a lot riding on its 1940 models. The marque's solid dependable reputation had been severly tested, the year before, when a redesigned and shortened frame had nearly proved disasterous. So, the 1940 Buicks offered more value for the car buyer's dollar than ever before.

For the first time, Buicks were offered in five distinct series, which shared the Special's 121-inch wheelbase (up an inch from 1939). The Supers filled the gap between the economical Specials and the larger, more powerful Centurys.

This bit of marketing strategy was a smashing success, as the new series 50 outsold all others. Buick retailed a record 310,995 cars, including the four millionth Buick of all time.

One big reason for this popularity was styling. The new "torpedo" bodies eliminated the runningboards. Combined with a smart new front end treatment that blended the headlight pods into the front fenders, Buick looked as modern and up to date as any car on the road.

Those headlights were now sealed beams. Mounted atop the pods were chromed parking lights which also served as directional signals, making Buick among the very first with that feature. Also, all Buick engines came equipped with oil filters.

Inside, the cars was an engine-turned dashboard, upgraded interior fabrics (with more color choices) and white plastic decorative trim in more places than ever before.

One of the affordable models that is popular with collectors today was the 1940 Buick Special Convertible Coupe, Model 46C.

SPECIFICATIONS

Year	1940
Make	Buick
Model	Special
Body style	Convertible Coupe
Base price	$1,138
Engine	OHV Inline Eight

Bore x stroke	3-3/32 x 4-1/8 inches
CID	248
Compression ratio	6.15:1 (Manual)
Carburetor	Carter AFB 4V
H.P.	115 @ 3500 rpm
Wheelbase	118 inches
Overall length	202-17/32 inches
Weight	3,780 pounds
Tires	6.50 x 15
OCPG Value	$36,000

Profile view shows the classic lines of the 1940 Buick convertible contrasted with classic architecture of a restored Wisconsin farm house. Buicks of this era offered more value for the car buyer's dollar than ever before.

The 1940 Buicks were introduced on September 22, 1939. Sealed beam headlights and new Fore-N-Aft Flash-Way directional signals were introduced.

The 1940 Buick front end ensemble was handsome. The headlights were sealed beams. Mounted directly above were parking lights which served as directional signals, making Buick among the first with them.

Buick's image changed in 1953

By Robert C. Ackerson

It was 1953 and the big Buicks, with their new V-8 power plant, had an exciting new performance image for a usually conservative nameplate. It was somewhat subtly packaged and promoted, but added go-power was there just the same. While the Roadmaster V-8 was touted as a car with "panther-like performance," Buick also described it as possessing power "that pours like a silken Niagara." Both descriptions were accurate.

The silk-smooth flow of power could be related to Buick's revamping of its Dynaflow transmission. For 1953, was promoted as Twin-Turbine Dynaflow. By adding a second turbine and a planetary gear set, Buick made possible a torque multiplication increase (from 2.25:1 to 2.45:1) of nearly 10 percent, as well as an even smoother delivery of power. When starting from a full stop, the first turbine delivered the driving torque, which was multiplied by the planetary gearing. As speed rose and the torque from the first turbine fin picked up, that from the second increased fairly rapidly. At highway cruising speed, the second turbine was delivering all the torque to the rear wheels while the other was free-wheeling with no distinct upshift.

Twin-Turbine Dynaflow power delivery was, said Buick, "utterly smooth." *Consumer Reports* (May 1953) concurred and explained that Dynaflow "is very smooth and quiet and much better than it was last year." Although Twin-Turbine Dynaflow (standard on Roadmasters; optional at $192 on Specials and Supers) did have a low range, which Buick regarded as an emergency low, its greatest shortcoming was the lack of a suitable intermediate driving range. *Motor Trend* (June 1953) praised Dynaflow's low range for providing "plenty of power under severe driving conditions" but concluded, "it is geared entirely too low for practical use in mountains." This also made a Twin Turbine Buick inferior to its competition in high-speed acceleration.

The sole example of Buick's straight eight heritage available in 1953 was found in the low-priced Special series. As it had for many years, the Buick Special offered Buick prestige and features at a very competitive price. The two-door sedan was Buick's price leader at $2,196.88. Other Special models were priced at $2,255.32 (four-door sedan); $2,295.43 (hardtop coupe) and $2,553.17 (convertible). Its engine, which had debuted back in 1950 as the Super's F-263 engine, displaced 263.6 cubic inches. It was rated at 125 horsepower and had a 7.0:1 compression ratio when linked to a manual transmission. The figures increased to 130 horsepower and 7.5:1 compression ratio when Dynaflow was installed.

With a road weight of 4,000 pounds, the Special's power-to-weight ratio of 32:1 promised leisurely performance that was born out by its 0-to-60 miles per hour time of 19-plus seconds.

Buick's torque tube drive and four wheel coil springs were intended to provide (in Buick's words) a "suave and gentle ride." Nonetheless, Floyd Clymer reported (*Popular Mechanics*, June 1953), "For a large American car, the Super handles well on corners." As had earlier been the case, there were plenty of gripes about Buick's power steering. No one was complaining about the ease it brought to wheeling a two-ton Buick down Main Street, but *Motor Trend* (June 1953) said, "it gave the impression of having no physical connection between the steering wheel and the front wheels."

Since the 1953 models represented the tail end of the Buick styling cycle that had begun in 1949, there were no dramatic changes awaiting their buyers. However, the 1953 models were easily identified by their XP300-derived headlight/parking light arrangement and twin bullet-shaped taillights. Buick's familiar bombsight hood ornament was lowered into a chrome valley on Super and Roadmaster models. On V-8 models, it was joined by a chrome "V." Less obvious was the replacement of Buick's old "open from either side" hood with a conventional rear-hinged unit. Far more serious, however, was the problem that befell some Roadmasters equipped with power brakes, when the rubber o-ring sealing the master cylinder from the vacuum cylinder failed. In a very short period of time, the brake fluid was pumped out of the brake lines and into the engine!

Buick quickly developed an improved o-ring and notified its dealers to replace the o-rings on all 1953 Buick Roadmasters brought into their shops. With wider 2-1/4-inch (up from 1-3/4-inch) brake drums on Special and Super models, Buick's overall braking performance was improved from 1952's level. For example, whereas a 1952 Roadmaster tested by *Motor Trend* needed 239 feet to stop from 60 miles per hour, a 1953 Super stopped from the same speed in just under 191 feet. Not all of this improved braking performance could be credited to the Buick's larger brakes, however. The 1952 Super model, at 4,530 pounds, outweighed the 1953 Super by approximately 230 pounds.

Also in 1953, General Motors described its Wildcat dream car (Buick's contribution to the 1953 Motorama show) as, "a single-seat sports convertible of futuristic design." Unlike the Corvette, the Wildcat wasn't destined for production, but its name was recycled and used both on future General Motors dream cars and (nearly 10 years later) on a good-looking, full-sized, Buick high-performance car.

In its original dream car form, the Wildcat had a 114-inch wheelbase and an overall length of 192 inches. Critics, who were quick to point out the foibles of American design and styling efforts, had a field day with the Wildcat's "Roto-Static" front wheel discs. They remained stationary as the wheels turned. But, the Wildcat also reflected, with somewhat greater objectivity, Buick's growing interest in building automobiles with improved roadability and overall performance.

For example, its front suspension had zero degree caster, vertical kingpins and direct-action shocks. At the rear, Buick's familiar coil springs featured new radius rods. The Wildcat's engine-transmission powertrain was pure-stock, with a 188-horsepower V-8 and Twin Turbine Dynaflow. However, its power steering operated through a fairly quick (15:1) overall ratio, instead of the normal 23:1.

Although honors as Buick's 7,000,000th production automobile went to a Roadmaster sedan, in June 1953, the Buick that really stood out as the flagship of its 50th anniversary fleet was the Skylark convertible. Like Wildcat, this was a name that would return to play a major role in Buick's modern history. In 1953, it was applied to a low-production (1,690 built) convertible that Buick said was "like the world of flight ... on wheels." Buick was off base in referring to the Skylark as "a six-passenger sports car," since its basic mechanics and suspension were pure stock Roadmaster components. It had the 121.5 inch wheelbase and a 4,300-pound road weight that were not compatable with road performance usually associated with small European sports cars.

However, if regarded as a low-volume and expensive ($4,596) Buick which combined sporty styling and the luxury features of American convertibles, the Skylark was an extremely interesting automobile. Its interior was offered with leather upholstery (available in four colors) and such features as Selectric radio with foot control, electric antenna and Twin-Turbine Dynaflow. Power steering and brakes were standard equipment for the Skylark.

The Skylark's steering wheel center proclaimed 1953 as Buick's 50th anniversary by depicting the silhouette of an antique Buick that was soon identified as a 1904 model by Buick historians. But, no doubt more satisfying to the ego of the Skylark's proud owner was the inscription (encircling the old Buick) that filled the blank in the horn button. It read, "Skylark customized for" followed by a space for the owner's signature.

No one had to peer into the Skylark's interior to get the message that it was a very special Buick. This was very nicely conveyed by a profile four inches lower than a standard Roadmaster convertible; the elimination of the fender portholes; and a very graceful side spear that nicely harmonized with the Skylark's fender and door lines. Although critics were quick to remind us that "wire wheels do not a sports car make," those on the Skylark were absolutely stunning. When the prototype Skylark appeared, in July 1952, it was fitted with Borrani wire wheels, but production models used 40-spoke chrome-plated Kelsey-Hayes wire wheels.

Buick promoted the Skylark as an automobile "especially styled for those who want exclusiveness, plus the complete modernity of Buick's Golden Anniversary automobiles." Beyond any doubt, it handsomely delivered on that claim. But, just ahead was a whole new generation of Buicks with sharp, trim styling, that featured the Skylark name returning, in glory, from the past. Like the original 1953 edition, they would boast an exciting new performance image.

SPECIFICATIONS

Year	1953
Make	Buick
Model	Roadmaster
Body style	Skylark Convertible
Base price	$5,000
Engine	OHV V-8
Bore x stroke	4.00 x 3.20 inches
CID	322
Compression ratio	8.5:1 (Manual)
Carburetor	Stromberg or Carter 4V
H.P.	188 @ 4,000 rpm
Wheelbase	121.5 inches
Overall length	207.6 inches
Weight	4,315 pounds
Tires	8.00 x 15
OCPG Value	$46,000

Typical Buick buyers were shown in this publicity photo highlighting the 1953 Model 72R Roadmaster four-door sedan. (Buick Motor Division photo)

The introduction of the first hardtop led to a sales spurt in the early 1950s. Demand continued strong for the 1953 Roadmaster Riviera coupe. (Buick Motor Division photo)

For the young-at-heart successful businessman there was the Roadmaster convertible. (Blackhawk Auto Collection)

A real image-changer for Buick was its limited-production Skylark "Anniversary Convertible" with styling patterned after that of European sports cars and California "kustoms." (Buick Motor Division photo)

This Century was a Special

By John A. Gunnell

In the 1930s, Buick created an extra-fast car called the Century. It had the company's biggest engine in its lightest chassis. The model's name suggested a 100 miles per hour top speed. It became known as the "bankers' hot rod" because of its upscale image combined with its high performance.

After World War II, the Century car-line was dropped from Buick's model charts. There was no need to have a performance car in the postwar "seller's market" since anything with wheels had a long line of customers waiting to buy it. However, as the 1950s dawned, cars sales began to slow down and the need to produce exciting and powerful cars arose once again.

The Century was revived in 1954 using the same concept last seen in 1942. The V-8 engine from the Roadmaster series was installed in the small Buick Special to make the new postwar Century. With 195 horsepower, the big power plant made the 122-inch wheelbase, under two ton Special a real bomb, Four body styles were available: sedan, Sport Coupe, convertible and station wagon. Nearly 82,000 sales were racked up by these Centurys.

Questing additional customers, Buick gave its 1955 models new styling and the Centurys gained a higher-compression 236-horsepower engine. The company reasoned, correctly, that additional buyers might be attracted by expanding the number of Century body styles. There were two new ones created in 1955.

This was easy to do, since the Special series already included one more model (a two-door sedan) and added an all-new 1955 four-door hardtop. The latter, called a Riviera Sedan, became a high-volume product in the high-performance Century series. It generated 55,088 sales as a Century. But the other new "Century" was a totally different story. It was anything, but a volume-seller. In fact, only 268 were made.

Why was this? It didn't make sense at first glance. The Special two-door sedan was quite popular with 41,557 assemblies in 1954 and 61,879 in 1955. So it seemed to make sense that it would also sell big numbers as a 1955 Century. However, there was a reason for the low number manufactured. The 1955 Century two-door sedans were built only for the California Highway Patrol, (CHP).

It seems that the CHP, at that time, was interested in having a small fleet of Class A traffic enforcement vehicles that could keep up with nearly anything on the road. The CHP conducted tests comparing the performance of six American cars that looked promising for such use. Buick had supplied the testers with a hybrid that took the Century concept even farther. This car featured the Special two-door sedan body (lightest of all Buicks) with the Century power plant. It was exceptionally fast.

As a hybrid, the test car was not available for purchase in the regular retail marketplace. Actually, it was not even a true Century. At the same time, it wasn't a Special either. The Special's two-door sedan body was used, but from the firewall forward, the patrol car was a Century. In the tough trials, this combination gave the best performance in standing-start acceleration and top speed of all models tested.

Two-door sedans were traditionally the lowest-priced model that automakers marketed. So, using this body for patrol cars had another advantage. It helped Buick come in as the low bidder for the CHP contract.

Facts about these cars, plus a photo, were printed in the April 1955 issue of *Highway Patrolman* magazine. It reported that the first of the cars rolled out of the CHP shops in Sacramento on March 2, 1955. "Grapevine talk about powerful new patrol cars was not idle gossip," noted the article. "The new cars are on the way."

The figure of 270 cars was mentioned in the headline of the 1955 article, but official Buick production records indicate that 268 Century two-door sedans were actually built. Another apparent discrepancy exists regarding the transmission attachments. *Highway Patrolman* said that half of the cars would have Dynaflow automatic transmissions, but historians believe that all of them were delivered with three-speed synchromesh gear boxes.

Jim Ashworth, an Orinda, California car collector spent 10 years looking for one of these rare cars. He remembered racing against one in 1955, when he was 16 years old. Actually, it was his father who was driving when a patrol car began following the Ashworths' 1955 Buick Century Riviera Sedan. Jim and his brother had added glass-pack mufflers to the car. The family was on the last leg of a cross-country trip on which their father had picked up two tickets for excessive noise.

Upon seeing the black-and-white Century two-door sedan behind them, the Ashworths were afraid that a third citation was coming from their home state. To their surprise, the uniformed driver rolled down his window and said "Let's race at the next light." It seems the highway patrolman had a bet going about whether their stick-shifted Centurys could outrun a similar car with Dynaflow. When the light changed, the red four-door hardtop pulled away faster, but the driver of the patrol car "popped" second gear and sped by.

For Jim, that drag race was a car nut's dream and ownership of a stick-shifted Century two-door sedan became part of it. However, Buick collectors were always trying to sell him Special two-door sedans and Century two-door hardtops. They insisted the automaker had never turned out any Century "post coupes."

Finally, Jim answered an advertisement placed by the wife of a man who had bought one of the police cars at a CHP auction in 1956. It had been stored in a barn until the man's death in the 1980s.

The old patrol car was basically sound, but some law enforcement equipment had been removed. The Orinda Police Department helped Ashworth locate an old General Electric two-way radio and spotlights with the correct mounting brackets. The collector now uses his ex-CHP 1955 Century two-door sedan to tow his vintage racing car. To avoid hassles, he carries a letter from the deputy commissioner of the CHP explaining the legalities of operating a retired patrol car on California highways. He has two large signs reading "Out of service since 1955" that must be placed in the rear side windows.

SPECIFICATIONS

Year	1955
Make	Buick
Model	Century
Body style	two-door sedan
Base price	n.a.
Engine	OHV V-8
Bore x stroke	4.0 x 3.2 inches
CID	322
Compression ratio	9.1:1
Carburetor	Carter 4V
H.P.	236 @ 4600 rpm
Wheelbase	122 inches
Overall length	206.7 inches
Weight	n.a.
Tires	7.60 x 15
Estimated Value	$15,000

The 1955 Century two-door sedans were built only for the California Highway Patrol. The Special body was used, but from the firewall forward, the patrol car was a Century. Using this body for patrol cars had another advantage. It helped Buick come in as the low bidder for the CHP contract.

Buick Motor Division featured some rather flamboyant tailfins on its 1959 and 1960 models. The 1959 Electra 225 convertible was used as a pace car at the year's Indianapolis 500-Mile Race. (IMSC photo)

1959 Buick Indy Pace Car

By Bill Siuru

Much has been written about the rear fin treatment on the distinctive 1959 Chevrolets. Most have been less than rave reviews. Often, it is forgotten that Chevrolet's sister divisions had some rather flamboyant fins on their 1959 and 1960 models.

Buick's were as wild as any of them and 1959 ushered in several other changes for the marque as well. But, first let's look at the fin issue.

The Buick's fins started at the front of the body, reaching a climax at the rear corner, then abruptly descending to meet the lower edge of the rear deck lid. Most of the fin line was accented by heavy chrome trim.

Although photos may make it appear that the fins rose in height as they approached the tail, it is really an optical illusion. When viewed from the side, the entire sweeping fin feature line is essentially level. Even though the 1959 Buick's fins are every bit as wild as those on Chevrolets, the Buick version appears to be better executed.

Buick went through a complete nameplate revision for its 1959 models. Gone were the familiar names like Special, Century, Super and Roadmaster, which had been around for decades. Now the lowest-priced Buicks were the LeSabres, the mid-priced cars were Invictas and the top-of-the-line models were Electras.

The plushest models were the Electra 225s. This huge car's designation came from its overall length, a whopping 225 inches. The LeSabres and Invictas were a "mere" 217.4 inches in length and the regular Electras measured 220.6 inches.

In overall styling, the 1959 models were totally redesigned from the 1958s. About the only thing retained was the grille treatment, with its individual chrome squares. And for 1959, there were fewer squares. One thing missing on the 1959 Buicks was the traditional ventiports (or "portholes") associated with Buicks for many years, until 1958.

One distinctive model offered in all series except the Electra 225, was the four-door hardtop with a flat roof and large wraparound rear window. All five General Motors marques offered this "greenhouse" design in 1959. By 1962, the greenhouse hardtop design was gone from all General Motors marques. Count the side windows and you'll see why this model is called a "four-window hardtop." Electra 225s used the larger General Motors body and had a conventional "six-window" four-door hardtop body that was shared with Cadillac.

The Electra 225s also came in four-door sedan and convertible coupe models. Needless to say, the big ragtop is a very desirable special interest car today. With a production total of 5,493 units, it was the third rarest 1959 Buick (about 50 fewer Invicta ragtops were built and some 260 fewer Invicta wagons). The survival rate of open cars is also low, of course.

All Electra 225s had extra-wide trim moldings, a massive Electra emblem on the front fenders, an Electra 225 chrome script ahead of the front wheel opening, an outside rear view mirror and Super Deluxe full wheel covers. Naturally, the interiors were plusher than on any other Buicks. The convertible, in fact, came with genuine leather upholstery, plus power window lifts and a power-operated convertible top.

In the engine department, the low-rung LeSabre was powered by the same 364-cubic-inch, 250-horsepower V-8 engine used in all 1958 Buicks with its compression ratio raised to 10.5:1. Other 1959 Buicks, including Electra 225s, were powered by a new 401-cubic-inch V-8 that produced 325 horsepower. This motor was an especially handy feature in the big "Duece-and-a-Quarter" models, which weighed from 4,562 pounds (for the convertible coupe) to 4,641 pounds (for the four-door hardtop).

A specially-lettered 1959 Buick Electra 225 convertible was given the honor of pacing the Indianapolis 500 that year. "Buick Official Pace Car" it said on the door. Over the rear wheel opening was the wording "May 30, 1959 Indianapolis 500 Mile Race." Sam Hanks, the 1957 race-winner, was behind the wheel of the race-starting ragtop.

Despite the attractiveness of its products and efforts like the Indy 500 Pace Car promotion, Buick ran a weak seventh in the annual American auto sales race in 1959. The radical new fin treatment did not help to bring customers flocking to Buick showrooms as they had earlier in the 1950s. Calendar year production saw a sustained slide. In 1957, the Buick total had been a healthy 407,271 units. For 1958, it dropped to 257,124. This trend continued in 1959, when 232,579 units were built.

SPECIFICATIONS

Year 1959
Make .... Buick
Model .... Electra 225
Body style .... convertible coupe
Base price $4,192
Engine .... OHV V-8
Bore x stroke .... 4.1875 x 3.64 inches
CID 401
Compression ratio 10.5:1
Carburetor .... Carter/Rochester 4V
H.P. .... 325 @ 4400 rpm
Wheelbase .... 126.3 inches
Overall length .... 225.4 inches
Weight .... 4,662 pounds
Tires .... 8.00 x 15
OCPG Value $24,000

Posing with the pace car. Buick's fins started at the front of the body, reaching a climax at the rear corner, then abruptly descended to meet the lower edge of the deck lid. (IMSC photo)

The plushest models were the Electra 225s. This huge car's designation came from its overall length, a whopping 225 inches. The convertible had a $4,192 list price and many were loaded with extra-cost options. (Buick photo)

A young lady with a stylish hat looks over the first Riviera. The 1963 model reintroduced the name formerly used to identify Buick hardtops on a new type of luxury Sport Coupe.

1963: The premier Buick Riviera

By R. Perry Zavitz

It was amid eager anticipation that the 1963 Buick Riviera greeted the public. This wasn't the very first Buick Riviera. Buick built a lot of Rivieras before 1963. They were hardtops, but they are not the cars we want to focus on.

Ford's four-passenger Thunderbird, of 1958, uncorked a whole new segment of the market, which, until then, no one really knew existed. It was a growing group of buyers who were comfortably well off. They wanted an agile car; not the monstrous sedans most manufacturers offered in this price range. These cars were about mid-size in today's lingo.

Compact cars were too cramped and lacked the performance these people were used to. Compacts also had a cheap aura about them which, of course, these buyers would not consider for a moment. Sports cars were too extreme. They certainly had agility and performance, but they were short on the comfort these people were accustomed to and unwilling to give up.

The four-seater Thunderbird was put together with the right combination of these desired characteristics. It had this lucrative market all to itself until the 1963 Buick Riviera appeared in showrooms all across the land.

Riviera styling was totally new and bore no resemblance to existing or previous Buicks. This could have been a very risky move, but it was a resounding success in the Riviera's case. No doubt the Riviera's great appeal was due in no small way to its pure and simple body lines. It came close to the simple elegance of such great designs as the 1936-1937 Cord and the Continental Mark II.

The unfamiliar new body covered a mixture of old and new Buick components. The 117-inch wheelbase came from Buick's X-type frame, which was a shortened version of the full-size X frame. Wheelbase was about five inches longer than the contemporary Buick Skylark, but overall length was almost 1-1/4 feet more. Both front and rear treads were two inches narrower than the full-size Buicks, but three to four inches wider than the mid-size models. So, Riviera had a totally different stance than its fellow Buicks.

Mechanically, the Riviera borrowed heavily from the senior Buicks. The engine was the 325 horsepower 401 cubic inch V-8 that powered the Invicta and the Electra. The transmission was Buick's Turbine Drive automatic.

The transmission lever was located on the console, between the bucket seats. The Riviera's interior was sumptuous, befitting a car bearing the Buick name. Elegant looking all-vinyl upholstery was standard. But, buyers had fabric-and-vinyl or leather-and-vinyl interiors to select from the option list, if they wished.

Other creature comforts, not on the option list but standard equipment, included: electric clock, trip mileage odometer, two-speed wipers with windshield washer, courtesy lights, deep-pile carpeting, front and rear bucket seats, foam rubber seat padding, deluxe wheelcovers, power steering, power brakes and many more items too numerous to mention.

Despite the long list of standard equipment, there was an option list, too. Items such as cruise control, air conditioning and wire wheelcovers could be ordered at extra cost. There was an optional engine as well. It was a variation of the standard engine, but with a one-eighth inch larger bore. That increased the displacement to 425 cubic inches. Consequently, the horsepower was boosted to 340.

Although it was smaller than the Electra and LeSabre, the Riviera tipped the scales at just about the same amount as the LeSabre two-door hardtop; a hair under two tons. But, its price was about $170 more than the Electra two-door hardtop. For a smaller car, its weight and higher cost was an indication that the Riviera was better equipped and more luxurious. Its base price was $4,333. Surprisingly, that was more than $100 under the lowest priced 1963 Thunderbird model. That states quite clearly who Buick was trying to lure into their sales offices.

As for performance, the new Riviera could do 0 to 60 miles per hour in 8.1 seconds and a quarter-mile in 16.01 seconds at 85.71 miles per hour. John Bond reported in *Car Life* a 0 to 60 time of 7.7 seconds with the 340-horsepower engine. Either engine gave the Riviera performance that was superior to the contemporary Thunderbird. These test results also substantiated the subtle look of performance the Riviera had.

Sales of Buick's new car were so good that an even 40,000 were produced for 1963. Meanwhile, Thunderbird sales dropped just over 13 percent during 1963.

Only a few minor details altered the 1964 Riviera's exterior appearance. The same 325-horsepower engine remained standard. The 340 horsepower engine was joined on the option list by a 360-horsepower motor. It used two four-barrel carburetors, instead of just one. Production was down slightly to 37,958 for the 1964 model.

The first noticeable styling change for the Riviera came on the 1965 model. Most obvious were the hidden headlights. They were located behind the grilles on the front end of the fenders. This change was an improvement, making the Riviera look even cleaner than before. Many people think the 1965 edition was the best-looking Riviera, at least until 1979.

Probably the most significant mechanical change made to the 1965 Riviera was the Gran Sport option. The GS package included the 360 horsepower engine, as well as dual exhausts and posi-traction rear axle with a 3.42 ratio (3.23 was standard). Heavy-duty suspension was not part of the GS option, but was available with or without it.

A totally new Riviera appeared for 1966, using a variation of the new front-wheel-drive Oldsmobile Toronado body. The Riviera stayed with rear-wheel-drive. Indeed, this Riviera has stayed with its 1963 basic concept of providing personal transportation for those wishing comfort, luxury and performance and who have no worries about meeting the payments. Riviera has frequently been Buick's best selling model during the last quarter century.

SPECIFICATIONS

Year	1963
Make	Buick
Model	Riviera
Body style	two-door hardtop
Base price	$4,333
Engine	OHV V-8
Bore x stroke	4.1875 x 3.64 inches
CID	401
Compression ratio	10.25:1
Carburetor	4V
H.P.	325 @ 4400 rpm
Wheelbase	117 inches
Overall length	208 inches
Weight	3,998 pounds
Tires	7.50 x 15
OCPG Value	$13,000

Front three-quarter view reveals the Riviera's handsome grille. Few significant styling alterations were seen on the 1964 edition, which retained the same crisp lines and design details. Note the 00-00-00 license plate.

The rear view of the Riviera featured "razor edge" fender lines and low-mounted horizontal rectangles for taillamps. Crisp shapes and a tailored look dominated the entire overall design.

The development of the second Cadillac V-16 was stimulated by the golden glow of a dying era of motoring glory. This 1940 V-16 Derham five-passenger coupe is from the Merle Norman Classic Beauty Collection of Sylmar, California.

Cadillac's last V-16

By Robert C. Ackerson

Most automotive historians agree that the seeds of Cadillac's tremendous postwar popularity were sown during the waning years of the 1930s.

Many things were in place by 1941. The LaSalle companion car was discontinued. Hydra-Matic transmission was made available. Styling elements that would be part of the Cadillac "look" for another two decades were established. Plus, a one-engine policy (that reduced costs without adversely affecting Cadillac's image) had been put into place.

The latter development sounded the death knell for one of the 1930s most exotic engine designs: the V-16. Cadillac had introduced its first V-16 series in early 1930. It was not the most propitious time for a very expensive automobile to begin its career. Initially, however, demand for the V-16 models was strong: 3,250 were produced in 1930. This popularity was a reflection of the appreciation that discriminating motorists had for the V-16 engine's capability. It not only delivered 165 horsepower at 3400 rpm and 320 pounds-feet of torque at 1200 rpm, but it did this with an ease that earned both the admiration and envy of Cadillac's competitors, domestic and foreign alike.

The next year Cadillac unveiled a V-12 running mate to the V-16. It displaced 370 cubic inches and was rated for 135 horsepower at 3400 rpm and 284 pounds-feet of torque at 1200 rpm. This meant that Cadillac was producing five different engines, since it also offered V-8-powered Cadillacs and LaSalles.

When Nicholas Dreystadt became Cadillac's general manager, in 1934, his goal was to cut costs without sacrificing quality. The proliferation of power plants had its days numbered. After a new V-8 was introduced in 1936, the V-12 was eliminated in 1937. In 1938 came the introduction of a less costly V-16. This engine had an almost unbelievably low production run. Only 514 were built from 1938 to 1940.

Such limited output makes it tempting to regard the second V-16 as little more than an aberration stimulated by the golden glow of a dying era of motoring glory. Yet, in many ways it represented Cadillac's grasp of sound engineering and manufacturing principles ... qualities the firm carried over into the fabulous 1940s and 1950s.

In discussing details of the V-16's development in *Automotive Industries* (November 27, 1937) Ernest W. Seaholm, Cadillac's chief engineer, outlined the design goals pursued by his staff in its creation. The plan was to design a motor having more than eight cylinders and as much power as the existing V-16. It was to be shorter in length than the V-12 and V-16; lighter than both; less expensive to produce; and able to perform to Cadillac's very high standards.

For a time a V-12 was considered, but the V-16 overwhelmed this V-12 in its ability to provide both smoother performance and, with a shorter piston stroke, greater durability. This left Cadillac's engineers to contend with the higher manufacturing costs associated with a V-16 engine. Seaholm was unperturbed, writing: "We believed, however, that we could develop a simplified design in which 16-cylinder advantages would be realized in an engine having actually fewer parts than either the Cadillac 16 or Cadillac 12 then in production."

Seaholm achieved not only this objective, but all the others on Cadillac's "dream engine" list. The second V-16 had 1,627 numbered parts, whereas the old V-16 had 3,273 and the V-12 had 2,810. A similar advantage was enjoyed by the new V-16 when its weight was compared to that of the superseded V-12 and V-16. While their weights were 1,165 pounds and 1,300 pounds respectively, the new V-16 weighed 1,050 pounds.

Seaholm's engine, with 135-degrees between cylinder banks and side valves, admittedly lacked the elegant appearance of the previous V-16 with its slender 45-degree V and overhead valves. However, the 135-degree design retained the equal firing intervals of the first V-16. It also lowered engine height (at the minor cost of greater width), resulting in a motor more compact than either the old V-12 or V-16. In terms of absolute power output, the second-generation V-16 developed 185 horsepower at 3600 rpm. Seaholm noted that this was, "Almost precisely the same as that of the former V-16."

Ultimately, the 16-cylinder engine became a victim of progress. Its L-head design possessed a limited ability to coexist with high compression ratios. A year before it even entered production, work was underway on Cadillac's overhead valve V-8 of 1949. Not surprisingly, the postwar V-8's "guiding light" was engineer Ed Cole, who would later gain performance fame at Chevrolet. Back in the early 1930s, Cole had been closely involved in the 135-degree V-16's development. In this case, at least a glorious past was very much a prelude to an even grander future.

SPECIFICATIONS

Year	1940
Make	Cadillac
Model	Series 90
Body style	five-passenger sedan
Base price	$5,695
Engine	L-head; V-16
Bore x stroke	3.25 x 3.25 inches
CID	431
Compression ratio	6.75:1
Carburetor	(2) Carter WDO
H.P.	185 @ 3600 rpm
Wheelbase	141 inches
Overall length	255-11/16 inches
Weight	5,140 pounds
Tires	7.50 x 16
OCPG Value	$82,000

Richie Clyne of Imperial Palace Auto Collection poses with W.C. Field's 1940 V-16. The 1940 Cadillac Series 90 V-16 Touring Sedan cost nearly $6,000 in standard form, without any extras.

In 1938 came the introduction of a less costly Cadillac V-16. Only 514 cars were built with this motor from 1938 to 1940. This photo shows a 1939 or 1940 Series 75 V-16 limousine in rear view.

A power seat and power windows and a special plush interior were standard for the first Coupe DeVille. (Old Cars photo)

A Cadillac coup: the 1949 DeVille

By Robert C. Ackerson

On October 22, 1948, Cadillac introduced the modern overhead valve V-8 engine to American motorists calling it the "greatest automobile engine ever built." Cadillac clearly recognized that this was a pivotal event marking the onset of a new era of performance and motoring excitement.

Like so many other automotive projects, it's probable that Cadillac's overhead valve V-8 would have reached the production stage far sooner than 1949, if not for World War II. In the mid-1930s, Cadillac was investigating alternatives to its existing line of V-8, V-12 and V-16 engines. It wanted a power plant that would be less expensive to produce, but one that would take advantage of the knowledge amassed by "Boss" Kettering's research into high-octane fuel development.

In describing this time span to prospective 1949 customers, Cadillac explained, "As far back as 1936, Cadillac engineers foresaw the direction automotive power should take toward lighter, more powerful, high-compression engines. Their own experience indicated that these engines should be of V-type ... from 1936 to late 1941, experiments were made with virtually all types of design. By Pearl Harbor, certain basic facts had been learned; several experimental models had been made. During the war, the project had to be shelved, but it was revived immediately after V-J day. In late 1945, the first test engines were undergoing block tests. Meanwhile, basic research, in the typical Cadillac tradition, was going forward. Out of 12 years of effort, America's finest automotive engine was developed and is now presented in the 1949 Cadillac."

Cadillac took every possible measure to make certain that customers would be satisfied with the new engine. Before the first 1949 models were produced, the V-8s accumulated over one million miles of road use. In one test, a production model V-8 was pitted against an experimental model in a 100-hour endurance test run at a wide-open throttle setting of 4,250 rpm. At the test's conclusion, there was no appreciable wear on any engine parts nor had there been any mechanical failure.

On four different occasions, Cadillac V-8s were subjected to a 25,000-mile test cycle at the General Motors Proving Grounds. It exposed them to every conceivable strain and hazard, including sustained high speed runs, pulling tests on grades as steep as 27 percent, runs over muddy roads and water baths.

At the conclusion, Cadillac reported: "Those who know best ... test drivers and experimental engine specialists ... say without reservation that the new Cadillac engine is more powerful, more durable, more economical than any stock engine ever built, including the great previous Cadillac engines."

At one point, prototype models had displaced just 309 cubic inches. If adopted for production, this would have been uncomfortably close in size to Oldsmobile's new 303-cubic-inch V-8. Instead, Cadillac went to 331 cubic inches, a size that would remain unchanged until 1956, when displacement was increased to 365 cubic inches.

Compared to the old 346-cubic-inch V-8, the new V-8 was superior in virtually every category. In terms of weight it was 699 pounds or 190 pounds lighter than the old engine. Its far smaller physical size released designers from many of the constraints imposed upon them by the bulk of older engines.

Although the overhead valve V-8 would eventually have a compression ratio of 10.5:1 and a displacement of 429 cubic inches in 1967 (its final year of production) it began with a mild ratio of 7.5:1. This was only marginally higher than the L-head's 7.25:1.

While there wasn't a dramatic distinction in the power ratings of the two engines, their specifications differed in many ways. The 1948 engine had a smaller 3.50-inch bore, a larger 4.50-inch stroke and 346 cubic inches of piston displacement. It had fewer main bearings (three), less horsepower (150 at 3400 rpm), and a lower torque rating (283 pounds-feet at 1600 rpm compared to 312 pounds-feet at 1800 rpm for the 1949 engine).

Cadillac's initial performance claims for its 1949 models were modest. The new model, it reported, would reach 80 miles per hour from rest in 30 seconds. In contrast, a typical low-priced American sedan, such as the 1949 Ford six-cylinder, needed 38.6 seconds. Compared to the 1948 model's 16.3-second time, the 160-horsepower Cadillac accelerated from 0 to 60 miles per hour in 13.4 seconds. This performance wasn't quite matched by the test of the 1949 Series 62 model by England's *The Motor*, which reported the following acceleration times in its March 2, 1950 issue:

Speed	Time
0-30 miles per hour	5.1 seconds
0-40 miles per hour	7.8 seconds
0-50 miles per hour	11.6 seconds
0-60 miles per hour	15.8 seconds
0-70 miles per hour	20.7 seconds
0-80 miles per hour	31.7 seconds
Standing start quarter-mile	20.0 seconds
Top speed	99.7 miles per hour

Nonetheless, *The Motor* was extremely impressed with the 1949 Cadillac, describing it as "a vehicle manifestly intended to cover long distances at a high speed ..." Recognizing the Series 62 sedan as one of the less expensive Cadillac models, *The Motor* concluded its analysis by noting, "Viewed in this light the car offers astonishing value for the money. It has a performance which few makes can rival, even fewer surpass, a general silence of running (including low wind noise), which many will consider unbeaten, and an ease in driving which must be a great asset when very long mileages are attempted."

A new model name in the Cadillac 62 series was the Coupe DeVille, a pillarless coupe or two-door hardtop. It arrived late in the year, around July, after a prototype version had caused quite a stir at GM's "Transportation Unlimited" show at the Waldorf Astoria in New York City.

Many people described the new model as a "hardtop convertible." Cadillac designated it model 49-6237-X. The 49 indicated 1949, the 62 stood for the series, 37 was the code for the new body style and X was the code for power seat and windows, which were standard equipment. It also had a special plush interior.

On November 25, 1949, Cadillac built its millionth car. A two-tone Coupe DeVille was selected and lettered-up to note the honor. It was one of just 2,150 two-door hardtops the company built that year.

The low production makes any 1949 Coupe DeVille a fairly rare collectible today. To date, some five million Coupe DeVilles and Sedan DeVilles have been sold, making these nameplates the best-selling models in Cadillac's long history of building great motor cars.

SPECIFICATIONS

Year	1949
Make	Cadillac
Model	Coupe DeVille
Body style	two-door hardtop
Base price	$3,497
Engine	OHV V-8
Bore x stroke	3.8175 x 3.65 inches
CID	331
Compression ratio	7.5:1
Carburetor	2V
H.P.	160 @ 3800 rpm
Wheelbase	126 inches
Overall length	214 inches
Weight	4,033 pounds
Tires	8.00 x 15
OCPG Value	$27,000

A new model name in the 1949 Cadillac 62 series was the Coupe DeVille, used on a pillarless coupe or two-door hardtop. (Cadillac photo)

The Coupe DeVille arrived in July 1949, after a prototype version had caused quite a stir at General Motors' "Transportation Unlimited" show. (Cadillac photo)

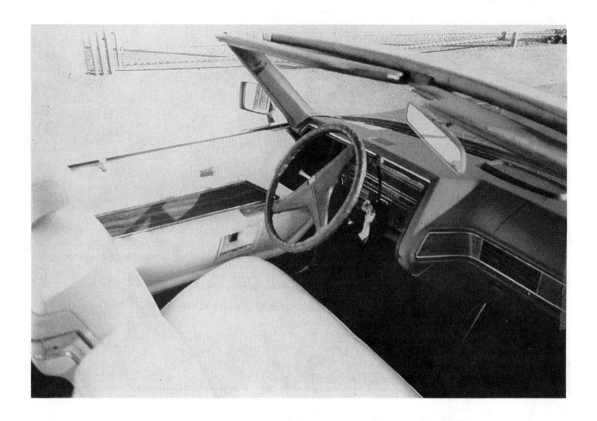

Restorers may also find that the extensive use of plastic parts inside the Convertible DeVille results in cracking of interior components exposed to heat and sunlight. This car had cracks on steering wheel and dashboard.

Cadillac's last 1960s ragtop

By John A. Gunnell

As every old car enthusiast knows, Cadillac convertibles of the 1950s have reached astronomical price levels at collector car auctions. This has led to increased demand and higher values for early 1960s ragtops bearing the Cadillac nameplate.

So, what about models of the late 1960s? How are they faring in the collectors' marketplace these days? And what do they offer the hobbyist seeking an affordable investment? To answer such questions, we have to take a closer look at cars like the 1969 Convertible DeVille.

Cadillac Motor Division built its four millionth car towards the end of the 1969 model-year. In all, 223,267 examples of America's favorite luxury car were produced that season, a 2.9 percent decline from the 1968 total. However, a big reason for the drop off was a United Auto Workers Union strike against General Motors. Product-wise, the Cadillacs that debuted on September 26, 1968 were every inch as attractive as the previous models.

The only open car available that fall was the model 6837F DeVille series convertible. It had a factory price of $5,887 and tipped the scales at 4,950 pounds with no extras. As experts will point out, the base price for the same car climbed to $5,905 after January 1, 1969, when front seat headrests, formerly optional, became federally-required safety equipment.

Like other 1969 Cadillacs, the convertible featured styling changes that weren't as extensive as those found on the year's Eldorado. However, refinements were made to give the Calais, DeVilles and Fleetwoods more of an Eldorado-like appearance.

The forward-thrusting front end of 1967 to 1968 DeVilles was replaced with a more squared-off profile. A new egg-crate grille with horizontal fins between its openings, was enhanced by a return to horizontally-mounted dual headlamps and wraparound parking/signal lamps. Both front and rear fenders had "razor edge" styling and a 2.5-inch longer hood added to a new impression of lowness and length. Also new was the use of one-piece side glass and the total elimination of cowl vents.

Technical changes ranged from the adoption of an improved ventilation system to the use of "guard rail" type safety beams in the doors. There was an improved energy-absorbing steering column, a new seat belt system, redesigned frame and a "closed" cooling system. The huge, powerful V-8 was carried over virtually unchanged from the previous season.

Standard equipment in the DeVille car-line included all the General Motors safety features, plus a V-8 engine; Turbo-Hydramatic transmission; variable-ratio power steering; dual power brake system; power windows; two-way power front seat with center armrest; rear armrest seat; electric clock; two front and rear cigarette lighters; twin front and rear ashtrays; courtesy, safety and warning lamps; mirror group; concealed three-speed windshield wipers; and 9.00 x 15 four-ply-rated black sidewall tires.

According to surveys of original 1969 Cadillac owners, they were pretty good cars. Answers to questionnaires indicated that they had a better than average service history and would travel many trouble-free miles with good care. A revised air conditioning system, though improved from the 1966 to 1968 type, was still a source of above-average repairs, however.

Otherwise, the cars were excellent products. The Cadillac engine, fuel system and brake system were all very highly rated by owners. They reported that far less than average repairs were needed in these areas. Better than average performance was also delivered by electrical components, the exhaust system and the Turbo-Hydramatic transmission.

Offering about average durability and service were Cadillac's body work, driveline parts and suspension components. These were still fairly heavy cars and this must have affected the longevity of some steering parts and the factory-equipment shock absorbers. Owners reported above-average repairs needed in these areas. But, overall, the marque had one of the best reliability records available from American automakers at this time.

Naturally, the 472-cubic-inch high-compression V-8 was a bit of a gas guzzler requiring a steady diet of leaded fuel for top performance. With its 10.5:1 compression ratio, the Cadillac collector who owns such a car today should consider lead additives or an engine rebuild with hardened valve inserts. Restorers may also find that the extensive use of plastic parts results in cracking of interior components exposed to heat and sunlight.

Cadillac built a total of 16,445 1969 Convertible DeVilles. Many of these fine cars survive today in fair to excellent condition. They are now starting to look like promising investments at the "affordable" end of the postwar car marketplace. But will their value ever catch up with those of the "tall tailfinned" 1950s Cadillacs? Only time will tell!

SPECIFICATIONS

Year 1969
Make .... Cadillac
Model .... DeVille
Body style .... two-door convertible
Base price $5,905
Engine .... OHV V-8
Bore x stroke .... 4.30 x 4.06 inches
CID 472
Compression ratio 10.5:1
Carburetor .... Rochester 4V
H.P. .... 375 @ 4400 rpm
Wheelbase .... 129.5 inches
Overall length .... 225 inches
Weight .... 4,590 pounds
Tires .... 9.00 x 15
OCPG Value $16,000

The forward-thrusting front of 1967-1968 was replaced in 1969 with this squared-off profile. The only open 1969 Cadillac was the Convertible DeVille. It had a price of $5,887 and tipped the scales at 4,950 pounds with no extras.

The shape of the vertical wedge-shaped taillights was repeated in the rear bumper ends. Cadillac built a total of 16,445 Convertible DeVilles in 1969.

The meaning of Marathon

By Jane A. Schuman

Checker Cab Manufacturing Company started building taxicabs way back in 1923. Beginning late in 1959 (1960 model year), the Kalamazoo, Michigan automaker added passenger car models. The Superba name appeared on these vehicles, which were non-commercial versions of the Checker cab designed for the "civilian" auto market. Both four-door sedans and four-door station wagons were offered in Standard Line and Special Line series.

The following model year (1961), Checker introduced the Marathon name. It was essentially the Superba Special with a catchier model designation. However, the Marathon models had a few extras like a modestly fancier grille and chrome side moldings.

All of the cars that Checker turned out were modified versions of its Model A8 taxicab. Therefore, both kinds of Checkers had very few year-to-year mechanical alterations. There were not many cosmetic changes either. In fact, they remained nearly the same throughout the 1960s and the 1970s.

Checker Superbas and Marathons are renowned for their durability and long-lasting endurance. Perhaps it's not a big surprise that *Webster's New World Dictionary* defines the word "Marathon" as a contest enduring over a long period of time. Obviously, the Marathon cars and station wagons were appropriately dubbed.

Former Checker owner Harry Ottens, of Valley Stream, New York, recalls buying a 1962 Checker Marathon station wagon brand new. He purchased the vehicle in the fall of 1961, paying $3,300 for it.

Otten's son accompanied him to the dealership to buy the Marathon. He was in the second grade at the time. "This car will last until the little fellow is in college," said the sales manager. He was right and then some. The Marathon survives today and Otten's son is nearly 40 years old!

That Checker station wagon is still being driven regularly, despite the fact that it has over 165,000 miles on its odometer. The only major work that has ever been done on the vehicle is a repaint and the addition of new upholstery.

All of this seems to be in line with 1962 Checker advertising, which bragged about the cars averaging over 150,000 miles of use with very little maintenance. The company's reputation was based on endurance, not on style or flashiness. The character of its product matched the definition of the word Marathon.

Although rather conservative looking, the Checkers offered amenities such as power steering for $64 extra and air conditioning for just $411. To equip cars using the standard three-speed manual transmission with overdrive cost $108 additional. Also offered was a choice of automatic transmissions. Buyers could get single-range automatic transmission for $222 or dual-range automatic transmission for $248. Another unusual technical option was an overhead valve cylinder head for the 226-cubic-inch six.

Many of the cars and station wagons that Checker built have stood the test of endurance and survived through many years of grueling use. Perhaps that's part of the reason that car collectors are gravitating towards them today. In addition, many of the individual models are quite rare.

The Marathon that Harry Ottens owns was one of just 1,230 passenger cars the firm turned out in 1962 and only a percentage of those were station wagons. Actually, that was pretty good for Checker, which also produced 8,173 cabs in the same season.

You almost have to wonder what would have happened to our "new-car-every-few-years" based economy if everyone had owned a 1962 Checker Marathon.

SPECIFICATIONS

Year	1962
Make	Checker
Model	Marathon
Body style	four-door station wagon
Base price	$3,004
Engine	L-head; inline; six-cylinder
Bore x stroke	3-5/16 x 4-3/8 inches
CID	226
Compression ratio	8.0:1
Carburetor	Zenith 1V
H.P.	122 @ 4000 rpm
Wheelbase	120 inches
Overall length	199.5 inches
Weight	3,720 pounds
Tires	6.70 x 15
OCPG Value	$7,600

Checker Marathons were modified taxis and had very few year-to-year mechanical alterations or cosmetic changes. (Colin Peck)

Checker Motor Company's reputation was based on endurance, not on style or flashiness. (Ward C. Morgan photo)

Many of the cars and station wagons that Checker built have stood the test of endurance and survived through many years of grueling use. (Checker photo)

You almost have to wonder what would have happened to our "new-car-every-few-years" based economy if everyone had owned a 1962 Checker Marathon. (Elliott Kahn photo)

Two-Tens and Bel Airs with Powerglide received a slightly more powerful version of the 235-cubic-inch Chevrolet "Stovebolt Six."

Expect wonderful things from 1953 Chevrolets

By John A. Gunnell

The attractive advertisement showed a 1953 Chevrolet climbing up a hill in San Francisco, California. "Expect these wonderful things from Chevrolet's new high-compression power," read the headline below the artist's rendering of the mid-series Deluxe 210 four-door sedan.

Greater acceleration, faster getaway and increased passing ability were the "wonderful things" that the advertising copywriters promised for this new engine, plus better fuel economy. "You can be sure of finer and more responsive performance under every driving condition," they advised. "But with all this, you might *NOT* expect greater gasoline mileage. Yet you get it."

The Deluxe 210 series was above the Special 150 series and below the Bel Air series in Chevrolet's pecking order. This was the first year of Chevrolet offering three car-lines. Previous models had been designated Stylelines or Fleetlines, which usually indicated the type of roof design (except in the case of station wagons and sedan deliveries). Basically, Stylelines were notchbacks and Fleetlines were fastbacks. Both styles been offered in basic Special or fancier Deluxe trim levels in the early 1950s.

However, for 1953, the series name indicated trim level rather than roof design. On the outside, the Special 150s had no body side moldings, rubber gravel guards and plain rocker panels. Bel Airs had side moldings with twin moldings on the rear fender, bright metal gravel guards and bright rocker panel moldings. In between was the Deluxe 210 with single moldings along the entire body, plus bright guards and rocker accents.

On the inside, the Special 150 had plain upholstery, rubber floor mats, a standard steering wheel and one sun visor. Bel Airs had two-tone vinyl-and-cloth trim, carpets, twin visors, a deluxe steering wheel and chrome roof bows in hardtops. Again in the middle was the Deluxe 210, with fancier cloth seats, carpets, dual visors and a one-step-up steering wheel that wasn't quite as fancy as the Bel Air's. It was a two-spoke design with a full horn ring, but did not have as much chrome trim or as fancy a center emblem.

Technically, there wasn't much difference in all three Chevrolet car-lines, except that Special 150s came only with the standard three-speed manual transmission. This gearbox was used only with a less sprightly 108-horsepower "Thrift-King" engine. This power plant was standard for all lines, but mandatory in Special 150s. Deluxe 210s and Bel Airs equipped with optional Powerglide automatic transmission received a slightly more powerful version of the 235-cubic-inch Chevrolet "Stovebolt Six." It was this 115-horsepower high-compression engine that was featured in the San Francisco hill-climbing advertisement.

The car in the ad, the Deluxe 210 four-door sedan was designated Model Number 2103 and Body Style Number 1069W. It was a full-size six-passenger model, built on the same platform used for all Chevrolets (except the new Corvette, of course). It was Chevrolet's bread-and-butter car and the winner in the "family transportation" sweepstakes. With 332,497 assemblies, it was far and away the year's most popular Chevrolet. The next closest was the 210 two-door sedan, of which 247,455 were manufactured. In third runner-up slot was the Bel Air four-door sedan. They built 246,284 of those.

While the 215-horsepower six was certainly not a major technological advance, promoting it heavily in Chevrolet advertising was very necessary. The handwriting was on the wall and the performance V-8 era was happening. Most automakers were already offering overhead valve V-8s. Even Ford had a new one coming in 1954 to replace its venerable flathead. The "Chevy" small-block wasn't quite ready, so the ad copywriters did all they could to make the Stovebolt Six sound exciting. It worked in 1953 and Chevrolet held its long-time number one sales position for another year!

SPECIFICATIONS

Year 1953
Make.. Chevrolet
Model .... Two-Ten (210)
Body style... four-door sedan
Base price.. $1,761
Powerglide Engine .... OHV; inline; six-cylinder
Bore x stroke... 3-9/16 x 3-15/16 inches
CID 235.5
Compression ratio... 7.5:1 (Manual)
Carburetor.. Rochester BC or Carter 2101S 1V
H.P. .... 115 @ 3600 rpm
Wheelbase...115 inches
Overall length ....195.5 inches
Weight .... 3,250 pounds
Tires .... 6.70 x 15
OCPG Value $7,400

The Brilliant New "Two-Ten" 4-Door Sedan.

Expect these wonderful things from Chevrolet's new high-compression power...

No matter *where* or *how* you drive, Chevrolet's new high-compression power brings you many wonderful advantages.

You probably expect greater acceleration. And it's yours. You enjoy faster getaway and increased passing ability.

You, no doubt, count on climbing hills with new ease. And you do.

You can be sure of finer and more responsive performance under every driving condition. But with all this, you might *not* expect greater gasoline mileage. Yet you get it.

One reason is the new 115-h.p. "Blue-Flame" engine. Teamed with the new Powerglide* automatic transmission, this new valve-

in-head engine delivers more power than any other engine in the low-price field.

Another reason — the greatly advanced, 108-h.p. "Thrift-King" engine. This highly improved valve-in-head engine brings the same advantages of more power and higher compression to gearshift models.

Your Chevrolet dealer will be more than happy to demonstrate *all* the wonderful things you will find in America's most popular car. ... Chevrolet Division of General Motors, Detroit 2, Michigan.

*Combination of Powerglide automatic transmission and 115-h.p. "Blue-Flame" engine optional on "Two-Ten" and Bel Air models at extra cost. (Continuation of standard equipment and trim illustrated is dependent on availability of material.)

MORE PEOPLE BUY CHEVROLETS THAN ANY OTHER CAR!

This attractive ad showed a 1953 Chevrolet climbing up a hill in San Francisco, California.

Foreground: Chevrolet's beautiful Two-Ten 4-Door Sedan. Background: The thrilling Bel Air Sport Coupe.

Why so many more people are buying Chevrolets than any other car...

It so often happens that new friends and new neighbors like those in our picture find a common meeting ground for conversation in their choice of a Chevrolet.

But that's not at all surprising when you consider that, year after year, more people *buy* Chevrolets than any other car. And this year the preference for Chevrolet is greater than ever. Latest available figures for 1953 show that over 200,000 more people have bought Chevrolets than the second-choice car!

The point is that the things *they* like about Chevrolet are things *you'll* like, too.

For example, they like Chevrolet's thrifty, spirited, high-compression power and beauti-

ful, durable Body by Fisher—and so will you.

They like the extra convenience of Chevrolet's responsive Powerglide automatic transmission* . . . the extra handling ease and driving safety offered by Chevrolet Power Steering*—and you'll agree.

And there's much, much more you'll like about Chevrolet—*the lowest priced line in the low-price field.*

Now's a good time to see your Chevrolet dealer. . . . Chevrolet Division of General Motors, Detroit 2, Michigan.

Optional at extra cost. Combination of 115-h.p. "Blue-Flame" engine and Powerglide available on "Two-Ten" and Bel Air models only. Power Steering available on all models.

MORE PEOPLE BUY CHEVROLETS THAN ANY OTHER CAR!

The 210 sedan seen in this 1953 advertisement was Chevrolet's bread-and-butter car and the winner in the "family transportation" sweepstakes with 332,497 assemblies.

Beautiful basics:
1958 Chevrolet Biscayne

By John A. Gunnell

Entertainers Dinah Shore and Pat Boone appeared on the cover of a folder promoting the "longer, lower ... lovelier by far" 1958 Chevrolets. The sporty new Impala was the main focus of attention in the color brochure, but other Chevrolets sure weren't ignored. Sporty Corvettes, glamorous Bel Airs and economical Delrays were depicted, along with a fourth model that seemed most likely to be the one to pop up in your neighborhood driveways. This was the work-a-day Chevrolet Biscayne.

Like all other 1958 Chevrolet model names, the Biscayne designation was new and further emphasized the sales theme "the only completely new car in its field." It was taken from Key Biscayne, Florida and applied to just two body styles. They had a rich, but not ostentatious, appearance.

Two- and four-door sedans were offered in Biscayne trim. They were identified by script plates at the front portion of the rear fender coves, wide sweepspears with white-painted inserts on the front three-quarters of the body, and a twin taillamp treatment on each side of the gull-wing rear end. Like Delray models, the Biscaynes came standard with small hubcaps, no sill moldings and no chrome doo-dads on the front fenders. They did have slightly upgraded interior appointments, as compared to Delrays, plus bright window moldings.

All 1958 Chevrolets shared a "dream car" image with dual headlamps, a glittery grille and heavily sculptured sheet metal. Rooflines were totally revamped and sedans featured a new thin pillar styling treatment that gave them the youthful look of a hardtop combined with (as Chevrolet put it) "the protective brawn of conventional sedans."

A larger new windshield with a wraparound appearance made it easier to see the USA in your Chevrolet. The same could be said of the full wraparound rear window, which greatly improved the driver's view behind the car.

Technical innovations for 1958 were as dramatic as the more modern visual impact. Along with an X-braced Safety-Girder frame, Chevrolet buyers found a full-coil-spring suspension that provided a soft floating ride on all types of road surfaces. Other mechanical updates included improved anti-dive braking for straight and level stops, feather-light ball-race steering, and an optional Level-Air suspension system.

A wider choice of engine and transmission choices than ever before was offered. The 235.5-cubic-inch/145-horsepower six-cylinder was standard and a 283-cubic-inch/185-horsepower small-block was the base Turbo-Fire V-8. A 230-horsepower Super Turbo-Fire version with high-compression cylinder heads and a four-barrel carburetor was available along with two (250- or 290-horsepower) Ram-Jet fuel-injected 283-cubic-inch V-8s. A big-block 348-cubic-inch Turbo-Thrust V-8 was available with 250 horsepower standard, plus in 280- and 315-horsepower Super Turbo-Thrust editions.

Three-speed manual transmission was standard with all engines. A new Turboglide automatic could be had with 348s or the least powerful fuel-injected engine. The old two-speed Powerglide was still around for sixes and non-fuel-injected 283s (which could also be had with manual overdrive transmissions). A close-ratio three-speed was optional with all engines. Apparently, a few cars were also special-ordered with Corvette four-speeds.

Biscaynes, of course, were primarily aimed at the bread-and-butter market. Three-on-the-tree gear shifting is commonly found in the cars of this line and many seem to come with sixes or two-barrel small-block V-8s under their hoods.

Anyone considering the purchase of a Biscayne as a hobby vehicle will likely find this model fine for show-and-go purposes. While sedans do not offer the high collectibility of open cars or sporty hardtops, they are fine for everyday use, parade duty, long-distance touring and occasional participation in shows. During 1987, when this article was written, only one Biscayne was recorded sold at a collector car auction. The number 2 condition two-door sedan brought $1,925 and the market hasn't gone any higher since then.

SPECIFICATIONS

Year	1958
Make	Chevrolet
Model	Biscayne
Body style	two-door sedan
Base price	$2,290
Engine	OHV V-8
Bore x stroke	3.56 x 3.94 inches
CID	235.5
Compression ratio	8.25:1

```
Carburetor ...................................................................................Rochester 2V no. 7012127
H.P. ................................................................................................... 145 @ 4200 rpm
Wheelbase ................................................................................................117.5 inches
Overall length ..........................................................................................209.1 inches
Weight ..................................................................................................3,447 pounds
Tires ..............................................................................................................7.50 x 14
OCPG Value ................................................................................................... $5,700
```

For 1958, Chevrolet offered two- and four-door sedans in bottom-level Biscayne trim. This is the two-door sedan with optional two-tone finish. (Chevrolet photo)

The Biscayne 2-Door Sedan with Body by Fisher and Safety Plate Glass all around.

YOU'LL LIKE BEING LOOKED AT *in your beautiful* **'58 CHEVROLET.** *Rightly so. For you know that the sculptured lines of that longer, lower Body by Fisher set a new style in styling. And every move your Chevy makes tells you there's new high-mettled performance to go with that exclusive high-styled look.*

There's a special kind of glow that goes with owning a new Chevrolet. Behind the wheel, you feel like you're right where you belong. You know you're being looked at—and you couldn't look better.

You're driving the car with the styling that's causing the year's biggest stir. The plain fact is, people like to look at Chevrolets. They especially like those boldly sculptured contours and that graceful gull-wing rear. You can't miss or mistake a Chevy!

But this car brings you satisfaction that goes far beyond its beauty. It surrounds you with the bank-vault solidity of famous Body by Fisher. It carries you serenely over the miles with a smoothness that could only come from a new kind of Full Coil suspension—or Chevy's real air ride*. It responds with a silken rush that tells you here's something wonderfully new in the way of V8 power.

Driving this new Chevrolet is much too good to put off. Your Chevrolet dealer will arrange it. . . . Chevrolet Division of General Motors, Detroit 2, Michigan. *Optional at extra cost.*

CHEVROLET

This ad shows the Biscayne's twin taillamp treatment. Like all other 1958 Chevrolet model names, the Biscayne designation was new and further emphasized the sales theme "the only completely new car in its field."

SWEET

SEVENTEEN —

AND NOT

TO BE

MISSED!

CHEVROLET

IMPALA CONVERTIBLE —another new luxury model in the Bel Air Series. How about those long, low lines! And colors— wait till you see the samples!

CORVETTE —America's only authentic sports car! Offers five spirited V8's, two with Fuel Injection;* three transmissions, including 4-speed manual shift.*

BEL AIR SPORT SEDAN —imagine getting this one with Turbo-Thrust V8* and Turboglide!* You'd have the smoothest power combination in Chevrolet's class.

NOMAD —star of a high-styled five-wagon lineup for '58! Seats for six in this one— and it surrounds you with luxury. Choose any Chevy engine; up to 280 h.p. in V8's.

BISCAYNE 4-DOOR SEDAN —in Chevrolet's middle priced series. Biscaynes, you'll notice, have a bright look of beauty that's all their own!

BISCAYNE 2-DOOR SEDAN —ready to take you for a super-smooth ride with Full Coil suspension at all 4 wheels! New air ride* is also offered.

BROOKWOOD 4-DOOR 9-PASSENGER —you can take half the neighborhood to school in this one! Upholstery is easy to keep clean; wears well, too.

DELRAY 2-DOOR SEDAN —you'll save with a Delray and still get everything Chevy's famous for; smooth ride, easy handling, real performance!

YEOMAN 4-DOOR 6-PASSENGER —comes with any one of Chevrolet's superb new engines. For biggest savings, choose the new 145-h.p. Blue-Flame 6.

YEOMAN 2-DOOR 6-PASSENGER —pile the family in and you're out for a good time; put in a cargo (up to a ½ ton of it) and you've got a willing worker!

*Optional at extra cost.

Here's Chevrolet's whole happy family. Here's styling that sets a new style—new developments in riding comfort that make the high-priced cars jealous—new peaks of performance (V8 or 6) in every model. Don't miss seeing and driving a '58 Chevrolet before you buy that new car. It's a beautiful way to be thrifty!... Chevrolet Division of General Motors, Detroit 2, Michigan.

Full-line ad shows Corvette, Impala, Bel Air, Biscayne, Delray and wagons. Technical innovations for 1958 were as dramatic as the modern visuals. They included an X-braced Safety-Girder frame and full-coil-spring suspension.

Easier Handling...Safer Going!

Quite a show—you behind the wheel of a smart and smooth '58 Chevy—with Chevrolet Power Steering. For it offers you a thrilling double feature—easier handling and safer going.

Test it yourself. Sit back and relax as you ease your way through tangled city traffic . . . or slip that big Chevy into a "postage stamp" parking place. It's easy, so easy—Chevrolet Power Steering saves up to 80% of turning effort!

Now aim that new Chevy out into the country. Round tight curves with confidence. Steer effortlessly over rutty roads or loose gravel with ease and safety . . . the same control that's ready to help you in any emergency.

Try it before you buy your next car. It's a new experience in relaxed and secure driving. Saginaw Steering Gear Division of General Motors Corporation, Saginaw, Michigan.

FORWARD [GM] FROM FIFTY

CHEVROLET POWER STEERING

Anyone considering the purchase of a Bel Air as a hobby vehicle will likely find this model fine for show-and-go purposes. This one was depicted in an advertisement at another type of show. It highlights power steering.

Chevrolet's 1961 styling used angular lines, aircraft-inspired sculpturing and (on Impalas) just enough bright work to boil the blood of an enthusiast. Special crossed flags grille badge seen here indicates the car has SS option.

Down-sized 1961 Chevrolet

By John A. Gunnell

"Trim new size" was Chevrolet's 1961 promotional theme. The year's Biscaynes, Bel Airs and Impalas had a 119-inch wheelbase and a 209.3-inch overall length. Sales brochures suggested that they were easier to park and better handling. However, the real reason the Chevrolets were smaller had little to do with driver convenience. The 1961 down-sizing was merely an ill-timed reaction to the first big wave of foreign car buying in America.

A 1958 recession had sent people looking for smaller, more economical cars. European manufacturers enthusiastically filled their needs. Detroit panicked. Designers and engineers rushed off to develop the first American compacts and re-think the proportions of standard-size automobiles.

For 1961, they had smaller conventional models ready to go to market. Unfortunately, the economy was back to full steam by then. Big cars reverted to high-demand status and the down-sized flagships of the fleet met a cool reception in the showrooms.

Collectors of today look at such cars as the 1961 Chevrolet Impala Sport Coupe in a different light than the buyers who shunned them initially. In addition to seeing a certain beauty of line in the trim-sized 1961 image, they've developed an appreciation of the scarcity and historical significance of these nearly 35-year-old Chevrolets. In addition, some like the unique combination of size, weight, styling and power that's unavailable in any other postwar Chevrolet performance car. The 1961 Impala Sport Coupe is a crisply-styled car that's seen a strong escalation of values within a relatively short period of time.

Chevrolet styling for 1961 was characterized by angular lines, aircraft-inspired sculpturing and (on Impalas) just enough bright work to boil the blood of an enthusiast. Sales catalogs described the overall appearance as "neatly tailored simplicity." Three distinct rooflines were highlighted.

Sport Coupes featured a roof/window treatment with gently sloping front pillars and sleekly angled rear feature lines. The clean, slim design had great structural strength and gave a delightfully light and airy "bubble window" effect.

A stylized rendition of the gull-wing look of 1959-1960 was continued at the rear of all 1961 Chevrolets. It looked best on Impalas, which added an extra taillamp on each side at the rear and tapering body side inserts painted a contrasting color. At the rear of this trim was a crossed racing flags emblem and model name-script. Impalas also provided an electric clock, brake warning lamp, back-up lights and deluxe wheelcovers as standard equipment.

Inside they featured seats done in pattern fabrics trimmed with soft-grain vinyl and finished with distinctive bright metal panels. There were six interior color combinations. Deluxe window cranks, fingertip door release buttons, extra-long armrests with built-in reflectors, foam rubber seat cushions, and rich deep-twist carpets were standard fare.

A special Impala steering wheel with a punched-out-for-competition look highlighted the new instrument console, which all 1961 Chevrolets sported. Its design put all important controls directly in front of the driver in a space-efficient layout. Higher seats and new wider doors were other selling points.

The Impala Sport Coupe sold for $2,597 in standard form with a 235-cubic-inch, 135-horsepower six-cylinder engine under the hood. A two-barrel "small-block" engine was the standard V-8. This motor also came with a four-barrel carburetor that added 60 extra horsepower. Optional were five different versions of the "big-block" 348-cubic-inch V-8 with 250, 280, 305, 340 and 350 horsepower. On the top of the heap (except for serious race drivers) was the 409-cubic-inch engine that generated 360 horsepower.

Introduced in 1961 was the first Super Sport option for the Impala. The sales catalog said that it was available to "personalize any model in the Impala series." It offered a choice of high-performance V-8s with a four-speed synchromesh manual transmission (305-horsepower engine also available with high-performance Powerglide) and included power steering, power brakes (with sintered metallic linings) and exclusive narrow band white sidewall tires.

An instrument panel assist bar, padded dash, electric tachometer and SS identification badges were featured as well. The stylish Super Sport also had full wheelcovers with simulated knock-off hubs and a heavy-duty suspension. The package was available for all body styles and a four-door hardtop with it was shown in sales literature. However, it's likely that most of the 142 Super Sports sold were Sport Coupes.

A 1968 survey of 100,000 Chevrolet owners indicated that less than average repairs were required on 1961 Chevrolets for camshaft, piston ring and engine valve problems. Average repairs were required on brakes, electrical systems, front ends, mufflers, rear axles, shock absorbers and transmissions. Above average repairs were reported for carburetors, body rattles, rust and wheel alignments.

SPECIFICATIONS

Year 1961
Make.. Chevrolet
Model ....Impala
Body style...two-door Sport Coupe
Base price.. $2,704
Base V-8 ....OHV V-8
Bore x stroke..3.875 x 3.00 inches
CID.. 283
Compression ratio.. 8.5:1
Carburetor..Rochester 2V
H.P. .... 170 @ 4200 rpm
Wheelbase...119 inches
Overall length ....209.3 inches
Weight ....3,480 pounds
Tires ....8.00 x 14
OCPG Value $12,000

Collectors of today look at such cars as the 1961 Chevrolet Impala Sport Coupe in a different light than the buyers who shunned them initially. The sporty and attractive "small" Chevys seem large today. (Chevrolet photo)

The 1961 Impala Sport Coupe is a crisply-styled car that's seen a strong escalation of values within a relatively short period of time. Dual horizontal headlamps and a simple grille characterized the front end.

The second-generation Corvette lasted only four model years, in contrast to the first generation's 10, but over 95,000 of them were built. This Sting Ray coupe belongs to Dave and Peg Lindsay.

Corvette introduces the Sting Ray

By R. Perry Zavitz

Convincing enthusiasts that the 1953 Corvette was a serious attempt to break into the sports car field was begun rather feebly by Chevrolet.

This plastic car came with a sluggish automatic transmission connected to Chevrolet's old stovebolt six (albeit souped-up to near Jaguar power levels). The $3,513 price was two-thirds more than the popular MG. Just 300 Corvettes were made, which also cast the manufacturer's seriousness in doubt.

It returned for 1954, but with no noticeable changes. At that time, how could the largest division of the world's biggest car maker dare offer a 1954 model looking like a 1953? Even the 1955 models appeared virtually unchanged, with one significant exception. Chevrolet now had a V-8 engine available in any of its models. The most potent 195 horsepower version was offered in the Corvette. That provided better performance, but the slushy Powerglide transmission was still mandatory.

Slowly, the idea that Chevrolet had something worthy of a little consideration for the sports car enthusiast was beginning to sink in. By 1956, a three-speed manual transmission was available. That cleared the way for more enthusiasts to think of the Corvette as a borderline sports car. Also, the six-cylinder engine was discarded and very powerful V-8 options were made available.

By 1958, whether or not they liked the Corvette, sports car purists had to recognize it as a home-grown sports car. But, that was half a decade after its introduction. It may not have achieved its due recognition that soon if the Ford had continued doing what so many people wanted; building the two-seat Thunderbird. The four-seat Thunderbird was the best thing that ever happened to the Corvette in its early years.

Now that the Corvette had no domestic competition in its class, it was able to comfortably make many changes that were hardly viable before. Minor styling changes were made each model year, yet the original 1953 shape remained evident. It lasted for 10 model years, which must have set some sort of American record for the post Model T era.

More and higher powered engine options were offered in the Corvette. Most were the same as available in Chevrolet passenger cars, but some with the greatest horsepower were obtainable only under the fiberglass hood of the Corvette.

The 10th anniversary Corvette was introduced with a totally new body. Well, it was 97 percent new. The lower edge of the rear was a carryover from the 1961-1962 Corvettes. But, the rest was new. Styling was largely based on a one-off Corvette experimental car of the late 1950s called the Sting Ray. The 1963 Corvette, also called Sting Ray, offered two body styles for the first time. Instead of just a roadster or convertible, a coupe was available as well. There was an auxiliary ($237) hardtop available for it. It was the coupe that caught the attention of many people.

New as the body was, it revived some extinct styling features. The long sloping back had been popular from the late 1930s to the early 1950s. The Corvette resurrected that shape, which became known as the fastback, and it has remained more or less over the last quarter century.

An odd feature the Corvette revived was hidden headlights. The only American mass-production cars that used them before were the 1936-1937 Cord, the Graham Hollywood and the 1942 DeSoto. Since the Corvette adopted hidden headlights in 1963, they have also made scattered appearances to the present time. But, the Corvette added two new twists to this idea. They were power-operated. They were also the first dual headlights to be hidden.

The 1963 Corvette body had one more characteristic that had appeared earlier. The trunk or stowage area had no exterior access. In other words, no trunk lid. This was a feature of some early slope-backed cars in the mid-1930s. The last example was probably the 1958 Metropolitan.

The mention of these previous features is not intended to degrade the 1963 Corvette at all. The coupe was a stunning car and carried all these features extremely well. It was both individualistic and up-to-date looking.

The fastback coupe, with its split window, is usually what we first think of when the 1963 Corvette is mentioned. There were 10,594 of them made. Actually, it was slightly out-produced by the 10,919 convertibles. Seldom thought of, outside the ranks of Corvette owners, is the convertible with the optional hardtop in place. Its notch back appearance is a bit unusual for a Corvette of this time.

Power for the 1963 Corvette came from a choice of four 327-cubic-inch engines. The standard motor was rated at 250 horsepower. Optional were 300- and 340-horsepower editions, costing $54 and $108 extra, respectively. All these had four-barrel carburetion. Most powerful, for $431 extra was a 360 horsepower fuel-injected engine.

Three-speed manual transmission was standard regardless of engine. A four-speed was optional with any engine. Powerglide was available only with the two least potent power plants.

By now, sports car devotees were more inclined to accept some luxury that previously was considered sacrilege in a sports car. Chevrolet, ever eager to meet those demands ... if not aggressively encourage their installation, offered power brakes ($43), power steering ($75), power windows ($59), tinted windows ($16), AM-FM radio ($174) and even air conditioning ($422). But, the purist was more interested in the heater/defroster deletion ($100 credit), posi-traction rear axle ($43), sintered metallic brakes ($38), four-speed transmission ($188), cast aluminum knock-off wheels ($323) and even the 36-gallon fuel tank ($202). While production was increasing to meet the ever-growing Corvette demand, 1963 model production shattered all previous records. There were 21,513 coupes and convertibles built, a jump of nearly 50 percent over 1962.

This second-generation Corvette lasted only four model years, in contrast to the first generation's 10. Yet, in the those four years, over 95,000 Corvettes were built. That was an increase of 38 percent over the first 10 years. The Corvette's shaky beginning was a thing of the past. No longer were people wondering if its history would be measured in months. It was now judged in generations. Is there any other car in American automotive history that has survived 40 years of economic boom and bust, yet remained faithful to its original concept as closely as Corvette?

SPECIFICATIONS

Year 1963
Make .... Corvette
Model .... Sting Ray
Body style .... two-door coupe
Base price $3,490
Engine .... OHV V-8
Bore x stroke .... 4.00 x 3.25 inches
CID 327
Compression ratio 10.5:1
Carburetor .... Carter AFB 4V
H.P. .... 250 @ 4400 rpm
Wheelbase .... 98 inches
Overall length .... 175.3 inches
Weight .... 2,859 pounds
Tires .... 6.70 x 15
OCPG Value $31,000

The revived fastback shape of the 1963 Corvette "split window" coupe made it the most distinctive American car of that year.

Even the 1964 convertible was a unique-looking Corvette with its hidden headlights. (Capitol Corvette & Classics)

Foreground: Impala Custom Coupe. Left: Impala Sport Coupe. Right: Caprice Sedan.

Try Chevrolet's silent ride of quality.

In 1968, Chevrolet was still the leading family car maker and the Caprice was the model that Chevrolet offered as proof. This advertisement shows Impala Custom Coupe in foreground, rear of Impala Sport Coupe and sleek lines of Caprice sedan.

1968 Caprice: Chevrolet luxury

By John Lee

Chevrolet's big news for the 1968 model year was totally new styling for the Corvette. The racy, low-slung design that replaced the beautiful Sting Ray after its successful five-year run really grabbed the fancy of the motor press. The new "Vette" showed up on several magazine covers and got lots more ink inside them.

Meanwhile, the Camaro ... introduced in 1967 ... was giving the Mustang a run for its money and the Chevelle SS-396 had some new competition. It was in the form of an economy-performance car called Road Runner that came from Plymouth.

With the performance explosion of the mid- to late-1960s, we sometimes forget that the automakers' main business was building family transportation. In that regard, Chevrolet was still the leader and the Caprice was the model that Chevrolet offered as proof.

Similar to the pattern established with the Bel Air and the Impala, Chevrolet introduced the Caprice name in 1965 as an option. RPO Z16 was available for the Impala four-door hardtop. It was principally a luxurious trim package with rich interior fabrics, body sill moldings, color-keyed striping on the sides of the body and black accents in the grille and rear panel. However, a stiffer frame and different suspension components were also included.

Again following past practice, Chevrolet made the Caprice a full series in 1966. It had a hardtop coupe, a hardtop sedan and six- and eight-passenger station wagons. Now, instead of moving up from a Bel Air to an Impala, the upwardly mobile had another step, from the Impala to the Caprice.

And 181,000 people took that step in 1966, not counting those who bought the top-of-the-line wagons. The line-up of body styles in the series would remain unchanged for the next several years. The novelty of the Caprice wore out somewhat, dropping the hardtop coupe and sedan production totals to 124,500 for 1967 and 115,500 for 1968, then rebounding to 167,000 for 1969.

For the 1968 model run, the Caprice was most easily distinguished from the more mundane Impala by its hide-away headlights. When closed, the headlight doors continued the rectangular pattern of the stamped grille out to the forward-thrusting fender tips. The slender bumper bar horizontally bisected the grillework and ended with ver-

103

tical tips on the fenders. The Caprice and other full-sized Chevys hid the windshield wipers away behind the aft lip of the hood for 1968.

A new rear bumper appeared to float between the deck and a smoother lower valance and taillights that had been inching downward the last few years were set into the bumper. Caprices and Impalas had three lenses on a side, with the center one a back-up lamp. (Bel Airs and Biscaynes had only two lenses per side).

Instead of the Impala's combination of a rub strip midway down the body side and rocker panel trim, the Caprice had a wide trim band just above the lower character line on the body. A protective rub strip could also be optionally added in the same location.

Chevrolets for 1968 showed up with a number of new safety features required by federal legislation. Armrests that shielded the inside door handles from being tripped accidentally, padded windshield pillars, energy-absorbing seat backs, front and rear seat belts and side marker lights were among them.

Other standard Caprice equipment included full wheelcovers, courtesy and ashtray lamps, an electric clock and an armrest in the center of the front seat. The four-door hardtop with basic equipment listed for $3,271, a premium of $408 over the equivalent model in the Impala series.

Unlike other full-sized Chevys, the Caprice was not available with a six-cylinder engine, but there was a full range of V-8s. One of the popular engines was a small-block based on the well-known 283-cubic-inch V-8 with its stroke increased from three-inches to 3-1/4-inches to provide 307 cubic inches. With a two-barrel Rochester carburetor it generated 200 horsepower at 4600 rpm. Various versions of the 327-cubic-inch "small-block" and 396-cubic-inch and 427-cubic-inch "big-block" V-8s were optional. Turbo-Hydramatic transmission, air conditioning, power steering and power brakes were now found in most Chevrolets.

Not everyone owned, or aspired to own, a performance car in 1968. Most families were completely satisfied with the reliable, comfortable and roomy sedans of the day. And for those for whom basic transportation wasn't good enough, Chevrolet offered the Caprices.

SPECIFICATIONS

Year 1968
Make.. Chevrolet
Model .... Caprice
Body style.. four-door hardtop sedan
Base price.. $3,271
Engine .... OHV V-8
Bore x stroke...3.875 x 3.25 inches
CID.. 307
Compression ratio... 10.0:1
Carburetor...Rochester 2V
H.P. .... 200 @ 4600 rpm
Wheelbase...119 inches
Overall length ....215 inches
Weight ....3,754 pounds
Tires ....8.25 x 14
OCPG Value $6,500

Caprice
The Grand Chevrolet
It can make you feel richer

1968 Caprice Sedan

It's the very top of Chevrolet's big car line—looks, feels and behaves it, too. Park it next to an expensive limousine—it comes off looking great! Compares just as well inside, with special Caprice Sedan touches like luxurious fabrics, fold-down center armrest, extra sound insulation, the look of hand-rubbed walnut and more. You can add concealed headlights, stereo tape system and other nice touches. Go ahead. Get into that limousine class at Chevrolet cost with Caprice. The best of two worlds. Be smart. Be sure. Buy now at your Chevrolet dealer's.

CHEVROLET

GM

For the 1968 model run, the Caprice was most easily distinguished from the more mundane Impala by its hideaway headlights. This advertisement described it as "the grand Chevrolet" and it was.

The only American car that gives you all these features at a popular price.

Caprice Coupe. Silent riding. Rich feeling. Its new roof line is classic. Nothing on the road is more distinctively stylish. New full glass windows—no ventipanes—let you see, and *be seen*, better. Plush new interiors with color-keyed deep-twist carpeting and the look of hand-rubbed walnut on the instrument panel and door panels. A new Astro Ventilation system. It brings in outside air without your having to open a window. Noise stays outside. Caprice. It *can* make you feel richer. At a Chevrolet price. Be smart! Be sure! Buy now at your Chevrolet dealer's.

Caprice
The Grand Chevrolet

Caprice Coupe

All Caprices had a wide trim band just above the lower character line on the body. This ad showed the exterior and interior of the 1968 Caprice Coupe and highlighted a long list of features at an affordable price.

1968 Nova Super Sport

By Tom LaMarre

Forget the joke about "Nova" meaning "no go" in Spanish. The durable Nova was a good car in its day and its long production run yielded collectible versions that are sleepers in the current market. Perhaps the best example is the scarce 1968 Nova SS.

Chevrolet advertised the first Nova as "The Not-Too-Small Car." Technically, it was the Chevy II Nova, but it more closely resembled a scaled-down version of the Chevelle than the original Chevy II. The 1968 Novas are easily distinguished from late-models by the "Chevy II" lettering along the top border of the grille.

The totally redesigned compact drew rave reviews. "It's a smart-looking car featuring more than a dash of sporting flavor," automotive editor Bill Kilpatrick wrote in the October 1967 issue of *Popular Mechanics*. "Available as the Nova coupe, a four-door and the Nova SS, the car is six inches longer than the '67s, boasts many styling features reminiscent of the more expensive cars.

"Power choices are many and transmissions include fully synchronized three- and four-speed manuals and an automatic. The Nova SS, with the 350-cubic-inch 295-horsepower V-8 is a dazzling performer. This model, by the way, comes with a lot of special performance-oriented trim. Bucket seats are available in the coupe only."

The SS option package (RPO L48) cost $210 and included a blacked-out grille with SS emblems, lower body trim with black center section and raised "Super Sport" lettering, twin louvered hood panels, sport steering wheel, 350 Turbo-Fire engine and red stripe tires mounted on six-inch rims.

Although Chevrolet Motor Division's press release pictured a Nova SS with plain wheelcovers, optional Rally wheels greatly improved the car's appearance. Fake mag-style wheelcovers were also available.

The effects of the new safety laws were evident in the Nova's side marker lights and shoulder harness, which were required on all cars sold after January 1, 1968. Other items in the standard safety package were windshield washers and defoggers, a dual cylinder brake system and impact-absorbing steering column.

The Nova appealed to a wide market because of its many engine choices. A 153-cubic-inch four-cylinder power plant was standard. The optional engines were a 230-cubic-inch six, a 307-cubic-inch V-8, plus 327- and 350-cubic-inch V-8s with four-barrel carburetors.

The Nova SS engine was derived from Chevrolet's original 265 cubic inch V-8 of 1955, instead of the up-to-the-minute SS-396 V-8. Increasing the 327-cubic-inch engine to 350-cubic-inches was simply a matter of stroking it, using a crankshaft with longer throws.

Chevrolet built almost 201,000 Novas in the 1968 model year, including an unspecified quantity equipped with the SS package.

SPECIFICATIONS

Year	1968
Make	Chevrolet
Model	Nova SS
Body style	two-door coupe
Base price	$3,490
Engine	OHV V-8
Bore x stroke	4.00 x 3.48 inches
CID	350
Compression ratio	10.25:1
Carburetor	Carter 4V
H.P.	295
Wheelbase	111 inches
Overall length	187.7 inches
Weight	2,915 pounds
Tires	8.00 x 14
OCPG Value	$6,000

This advertisement pokes fun at the long hood/short deck phrase used so much in 1960s auto writing. The 1968 Nova SS, with the 350-cubic-inch 295 horsepower V-8 was a dazzling performer and came with a lot of special performance-oriented trim.

Here, the same photo is used in another advertisement. The Nova SS engine was derived from the 265-cubic-inch small-block V-8. It was the later 327 cubic inch version stroked to 350 cubic inches.

The 1968 Novas are easily distinguished from late-models by the "Chevy II" lettering above the right-hand taillight. This catalog art shows the rear view of the SS model with blacked-out rear panel and mags.

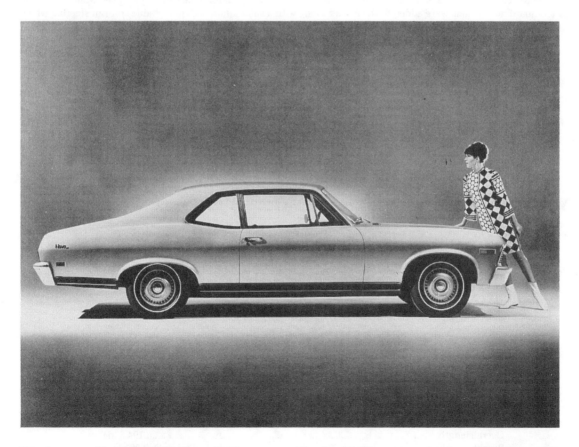

Handsome profile of the 1968 Chevy II Nova coupe. The durable Nova was a good car in its day and its long production run yielded collectible versions that are sleepers in the current market. (Chevrolet photo)

1973 Nova "hatched" new idea

By John A. Gunnell

Chevrolet "hatched up" a couple of changes in its 1973 Nova. Chief among them was a new five-door coupe, better known as a hatchback. Sales catalogs highlighted the model's fold-down rear seat and mammoth cargo capacity. "It's a practical car for people who don't quite need a wagon," a brochure said.

With the rear seat folded down, the Nova hatchback gave over six feet of cargo length and 27.3 cubic feet of load space. But, even though it was extra-functional, the car could be made sporty-looking. Unlike most station wagons, it had sleek lines and offered options such as bucket seats, a floor-mounted gear shift lever and the Super Sport equipment package.

The SS equipment group wasn't quite what it used to be in the high-performance years. However, with a low price of $123 extra, it was more attractive and popular than ever. The option could be ordered for either the coupe or hatchback in both the Nova and Nova Custom series. A total of 35,542 buyers specified the SS package for their 1973 Novas, although there is no record of how many cars of each body style and trim level were so-equipped.

A neat package for a young car collector would consist of the base 1973 Nova hatchback with the optional Turbo-Fire 307 engine. In standard form, the hatchback with this base V-8 was $2,618. A black vinyl top (needed to offset sexy colors like Medium Orange Metallic) was $83 extra.

Although the Nova hadn't really changed very much since 1968, Chevrolet had made several important improvements in the car almost every year. In addition to the new-for-1973 hatchback body, there was a new large-grid grille insert, rectangular dual-unit taillamps, re-engineered front and rear suspension systems, improved seats and sound insulation and new, forward-mounted, rectangular outside rear view mirrors.

On the engineering end of things, the 1973 bumpers had tougher reinforcements and improved front mountings. Side-Guard door beams were added to the Novas, along with a safer and more comfortable soft-rim steering wheel. One-piece door glass was another feature, along with Flow-Thru ventilation.

With the Nova SS package, the grille and rear panel were black-accented. SS emblems were placed on the grille, fender and rear deck lid, replacing Nova name-scripts in some locations. The parking lights and taillamps had bright accents, too. Dual Sport Mirrors (left-hand remote-controlled) were included, along with bright roof drip moldings, 14 x 6-inch Rally wheels (with special center caps and chromed lug nuts) and heavy-duty suspension parts.

Inside, an all-vinyl interior was seen. Bucket seats were optional in SS coupes and hatchbacks. Color-keyed deep-twist carpeting was part and parcel of the SS option and SS steering wheel emblems were included, too.

The 307-cubic-inch small-block V-8 came with a standard three-speed manual transmission. Turbo-Hydramatic was available with this engine, but a four-speed on the floor was not offered. (That came only with the four-barrel 350-cubic-inch engine). SAE net horsepower for the 307-cubic-inch engine was 115, which meant it wasn't good for serious drag racing. However, the Nova's size (111-inch wheelbase and 195.1 inches long) made the base V-8 adequate for most driving needs.

Is the 1973 Nova a collector car of the future? If you ask someone under 25 years old, they are likely to say yes. The Chevy II/Nova was always the kind of "wheels" that appealed to young buyers. Enthusiasts who drove these cars in their teens find nostalgia in them today.

SPECIFICATIONS

Year	1973
Make	Chevrolet
Model	Nova SS
Body style	two-door hatchback
Base price	$2,618
Base V-8	OHV V-8
Bore x stroke	3.87 x 3.25 inches
CID	307
Compression ratio	8.5:1
Carburetor	2V
H.P.	115 @ 3600 rpm
Wheelbase	111 inches
Overall length	194.3 inches
Weight	3,281 pounds
Tires	F78-14
OCPG Value	$4,000

There was a big SS badge on the right-hand corner of the rear deck lid. On the engineering end of things, 1973 front and rear bumpers had tougher reinforcements and improved mountings were used in front.

Chevrolet "hatched up" a couple of changes in its 1973 Nova. Chief among them was a new three-door coupe, better known as a hatchback. The new body style was indicated by lettering on the sail panels. Vinyl top was extra.

The SS equipment group wasn't quite what it used to be in the high-performance years. Standard under the hood of the Nova SS was the 307-cubic-inch small-block Chevrolet V-8.

Standard tires for the Nova Super Sport were size F78-14. The bright-finished Rally style wheel rims set off the raised white letters on the sidewalls. The side marker lamps had bright accents, too.

The 300 name was derived from its special motor. Bob Rodger dipped into the racing parts to create a production car as exciting as two-seaters. The cars were competitive on the old Daytona Beach Race Course. (Ray Doern photo)

The car that swept Daytona: 1955 Chrysler 300

By John A. Gunnell

Disc brakes, roller tappet cams and Air-Lift shocks on a 1954 Chrysler factory job? You bet your blue suede shoes! All this and more was a prelude to the Chrysler 300 letter car, America's original muscle machine.

The original edition was turned out as a mid-1955 model called the C-300. It was an image car designed to do battle with the Corvette and the Thunderbird. Chrysler wasn't in good enough financial shape to field its own two-seat sports car to stay competitive. Management had no idea how to put up a fight with Chevrolet and Ford, but MoPar's chief engineer Bob Rodger did.

Rodger liked racing and Chrysler had become a power house in several diverse fields of motorsports since 1951. That was the year Chrysler introduced a 180-horsepower hemi V-8 that was 12 percent more powerful than the then "hot" Cadillac V-8, even though both engines share the same displacement.

In no time at all, Chrysler hemi-powered stockers were competing in circle track, straight line and road racing contests. They took significant victories at Daytona Beach, as well as in the Mexican Road Race. At the same time, sportsman Briggs Cunningham campaigned his own Chrysler-powered sports car at LeMans.

Chrysler was aware that involvement in racing was good for sales of new cars and released many over-the-counter packages to help the racers. For example, Carl Kiekhaefer, who built Mercury outboard boat motors in Fond du Lac, Wisconsin, was sponsoring a team of Chrysler sedans in the Mexican Road Race. MoPar helped with installing two-barrel Zenith carburetors, Mallory ignition parts, special spark plugs, stiffer springs, telescopic shocks, oversize fuel tanks, heavy-duty front and rear sway bars and brakes that were cooled by dry ice.

The more Chrysler wins, the more factory racing options the company released. For 1953, about 15 cars were fitted with dual-quad induction, Air-Lift suspension, heavy-duty axles, Imperial disc brakes and 280-degree roller cams. These special 235-horsepower jobs were designed for NASCAR drivers.

By 1954, the Chryslers were dominating all kinds of racing, but they pushed the other factory-supported racing teams into taking defensive action. NASCAR rules were suddenly changed to limit stock racing cars to single four-barrel carburetors. This was okay with MoPar engineers, though. They simply bolted on an electric fuel pump, made a few adjustments and got the same 235 horsepower with just one carburetor. Then they got busy on another project.

This project was known as the Chrysler New Yorker "Export Special." The car was designed to compete in the Mexican Road Race. It had all of the NASCAR goodies and then some. When the race south of the border was canceled, Chrysler supplied such a motor to speed ace Tony Bettenhausen. With a train load of auto magazine and newspaper writers watching, the car circled the company's brand new Chelsa Proving Grounds at 118 miles per hour for 18 hours straight. It was said to have exactly 300 horsepower!

The name Chrysler 300 was derived from this special motor. Bob Rodger convinced the company to let him dip into the racing parts bin to create a production car as exciting as competing automakers' two-seaters.

The 1955 Chrysler C-300 was no Corvette or Thunderbird. In fact, it was more car than both put together. Designed for limited production only, it was a full-size model with luxury appointments, but it had a racing-type engine under its hood. Standard equipment included the special hemi; heavy-duty suspension; PowerFlite automatic transmission; windshield washers; custom steering wheel; dual exhausts; padded dash; power brakes; and custom tan leather interior. Introduced as a midyear model on February 10, 1955, the C-300 had a base price of $4,055.25.

Only 10 options, three exterior colors and one style of upholstery were offered. Buyers could add power steering, windows and seat; two kinds of radios; a heater; tinted glass; an electric clock; one of four axle ratios; and real Kelsey-Hayes wire wheels ($617.50 for the set of five wheels). Exterior color choices were Black, Tango Red or Platinum White. Seats were done in tan leather. Very early cars were mostly hand-built and some even had lopsided trim.

The real attraction was the hemi engine. It was a big-displacement block with relatively high compression for its day. Breathing was via twin WCFB carburetors and there was a dual-point distributor, over-sized exhaust system and solid valve lifters. The car could move from 0 to 60 miles per hour in around 10 seconds. The quarter-mile could be done in about 18 seconds with the car moving at 82 miles per hour at the end.

Chrysler 300 letter cars weren't meant for drag racing; they were tuned for top speed and high-end performance. Although they weighed over 4,000 pounds, the 300s would still be accelerating when they hit 100 miles per hour. The speedometer even had 150 miles per hour marked on it. In the Daytona "Flying Mile," two cars hit speeds of 126.54 and 127.58 miles per hour, respectively. They were the fastest production cars in the world at the time.

The C-300 became known as "the car that swept Daytona." Stock trim versions took first and second places in the beach racing that year and dominated both NASCAR and AAA competition. Driver Tim Flock won 13 races in a row with a car sponsored by Kiekhaefer's Mercury Outboard Racing Team. It was only the start of the Chrysler 300 story.

SPECIFICATIONS

Year	1955
Make	Chrysler
Model	C-300
Body style	two-door hardtop
Base price	$3,490
Engine	OHV V-8
Bore x stroke	3-13/16 x 3-5/8 inches
CID	331
Compression ratio	8.5:1
Carburetor	(2) Carter WCBF 4Vs
H.P.	300 @ 5200 rpm
Wheelbase	126 inches
Overall length	218.8 inches
Weight	4,005 pounds
Tires	8.20 x 15
OCPG Value	$36,000

Exterior color choices were Black, Tango Red or Platinum White. Seats were done in tan leather. Very early cars were mostly hand-built and some even had lopsided trim. (Old Cars photo)

The 1955 Chrysler C-300 was no Corvette or Thunderbird. The "typical" buyers shown in this contemporary publicity still were probably shocked by the rough, competition-like idle of the hemi engine. (Chrysler Historical)

Chrysler 300B took 16 wins in a row

By John A. Gunnell

The 1956 Chrysler 300B introduced the letter car series' first engine and transmission options. One new option was a high-horsepower hemi that helped a handful of the hot muscle cars win a string of stock car racing victories.

As far as the "standard" 1956 package, the base motor was bored to 354-cubic-inches. It now had milled heads giving a 9.0:1 compression ratio and 340 horsepower at 5200 rpm. The extra-horsepower job was announced on April 16, 1956, but was not released in production until June 13. It was the first American car engine option to offer one horsepower per cubic inch. However, since production was very limited, Chevrolet later claimed this honor for its "fuelie" 283 engine.

With the Chrysler 300B engine option, an honest 355-horsepower rating was the result, reportedly, of switching to a thinner head gasket, thus raising compression to 10.0:1. This engine was designed strictly for racing use and came with three-inch-diameter exhaust headers, which were also made optional for the 340-horsepower motor. Both blocks had hardened cranks, malleable iron bearing caps and tri-metal main bearings.

This was the only year that three transmissions went into letter cars, though not all at once. Early production units with automatic used the two-speed PowerFlite type. Later, the better three-speed TorqueFlite was substituted. Thirty-one cars had three-speed manual gear boxes with Dodge gear shift linkage and blocker plates covering the regular automatic push-button controls. They also had 1953 Chrysler steering wheels that were compatible with a stick shift. Power brakes were deleted from the standard equipment list for cars with stick shift.

The Chrysler 300B had integral-tailfin-style rear fenders, a more massive rear bumper and revised taillights. Otherwise, it was about the same as the 1955 C-300. At some point, Chrysler started painting the headlamp surrounds in body colors. They were formerly chrome-plated. As in 1955, the letter car's introduction was delayed. It debuted January 4, 1956 at a price of $4,312.25.

As in 1955, only three colors were offered. They were now Black, Regimental Red and Cloud White. The options list was extended to include air conditioning, an under-dash record player, a self-winding clock that was embedded in the steering wheel, and low-restrictive paper element air cleaners. Special 15 x 9.5-inch steel disc wheels were new and preferred by professional racers.

At the tracks, the 300Bs again swept Daytona, averaging 90.83 miles per hour on the Grand National beach oval course. Tim Flock hit 139.54 miles per hour in the "Flying Mile." The Mercury Outboard Racing Team took 16 wins in a row in NASCAR. They were invincible!

SPECIFICATIONS

Year	1956
Make	Chrysler
Model	300B
Body style	two-door hardtop
Base price	$4,312
Engine	OHV V-8
Bore x stroke	3.94 x 3.63 inches
CID	354
Compression ratio	9.0:1
Carburetor	(2) Carter WCBF 4Vs
H.P.	340 @ 5200 rpm
Wheelbase	126 inches
Overall length	221 inches
Weight	4,242 pounds
Tires	8.20 x 15
OCPG Value	$36,000

The 1956 Chrysler 300B introduced the letter car series' first engine and transmission options.

The Chrysler 300B had integral-tailfin-style rear fenders, a more massive rear bumper and revised tail-lights. (Chrysler Historical)

As in 1955, only three colors were offered. They were now Black, Regimental Red and Cloud White.

As far as the "standard" 1956 package, the base motor was bored to 354 cubic inches. It now had milled heads giving a 9.0:1 compression ratio and 340 horsepower at 5200 rpm.

The world's most complete car?

By R. Perry Zavitz

The 1963 Chrysler body was redesigned. While offensive fins were eliminated in 1962, the 1963 models looked like they never had such appendages. Still, the trapezoid-shaped grille shape clearly said this was a Chrysler.

Newports featured a grille with many thin, closely-spaced vertical elements. There were three wide-spaced horizontal elements as well.

The Sport 300 line had a grille with dull finish, except on the rim, and twin cross-bars, one vertical and one horizontal. Where these bars intersected, a circular red, white and blue medallion identified the series. This was almost the same as those worn by Chrysler 300 letter cars since 1957, but without a letter indicated. A Sport 300 convertible paced the Indianapolis 500, instead of the much more appropriate 300J.

That 1963 letter car did not follow the alphabet, as the previous ones had. The 1962 model was designated the 300H, but the 300I designation was never used. The manufacturer did not want buyers to misread the name as "Chrysler 3001."

Of great confusion was the nearly identical appearance of the non-lettered Sport 300 two-door hardtop with the 300J, which came as a hardtop only. The grille was quite similar, except that the medallion of the muscle car had a J suffix. There was also a squarish steering wheel inside, leather upholstery and a center console.

Chrysler's top car-line (not counting Imperial) was the New Yorker. Its grille was somewhat reminiscent of the 1955-1958 Imperials and 300 letter cars. A three-crown medallion was seen in the center of the grille cavity. But, on either side there was a grille section with powerful and aggressive-looking horizontal and vertical bars.

The New Yorker offered four-door sedans and hardtops, as well as six- or nine-passenger Town & Country station wagons. There were no two-door hardtops or convertibles in the New Yorker series at this time. In mid-season, a special option was offered on the four-door hardtop. It was called the Salon. The name was spelled out in chrome block letters on the front fenders. It clashed with the large gold New Yorker script directly above.

The Salon was a package that included a very generous list of options that we now expect on most luxury cars: AM/FM radio with rear speaker; cruise control; six-way power front seat; right front seat recliner; front and rear armrests; leather and cloth upholstery; color-keyed seat belts; deluxe carpeting and headlining; power brakes; power steering; power windows; TorqueFlite automatic transmission; air conditioning; windshield washers; Vari-Speed wipers; tinted glass; rear window defogger; padded vinyl roof panel; body side striping; and color-keyed wheelcovers. And the list went on and on. "Forty-four luxury items that you don't have to specify. They are yours automatically in the Chrysler New Yorker Salon," the brochure boasted.

Of course, all that extra was not exactly given away. The Salon listed for $5,344 ... practically double the price of the Newport sedan and $1,021 more than the base New Yorker hardtop. In case you forgot what a dollar bought 30 years ago, it should be pointed out that the New Yorker Salon cost more than any Cadillac 62 Series model, except the convertible. A base New Yorker four-door hardtop ($4,323) was priced almost as high as a Buick Electra 225 four-door hardtop ($4,448).

Mechanically, the Salon was no different from the other New Yorkers. Chrysler's 413 cubic-inch V-8 rated at 340 horsepower motivated this 4,300-pound car. (Its weight was 340 pounds higher than the base New Yorker, which is a good indication of the load of extras the Salon carried). The New Yorker had no engine options and could not be standard-ordered with the special high-performance version of the 413 offered in the 300J.

Like all 1963 Chryslers, the Salon had a 122-inch wheelbase. Overall length was 215-1/2 inches. Both wheelbase and total length were shortened about four inches from the 1962 New Yorker's measurements.

Due to both its mid-season introduction and very hefty price tag, there were only 593 Salons built. That was not many more than the 400 Chrysler 300Js produced that year.

The Salon made a return appearance for 1964. Except for re-shaped taillights and a revised grille (a bland affair with a multitude of thin horizontal bars) there was not very much different about the 1964 Salon. Despite a token price hike of $10 on the regular New Yorker hardtop, the Salon's price was up to $5,860. However, the package now contained 38 items, instead of the original 44.

Neither car size nor engine power were changed from 1963 for the Salon or other New Yorkers. Because it was now better known and had a full season of production in 1964, there were more than two-and-one-half times as many produced. The total was 1,621 in all. But, the two-year total was only 2,214.

SPECIFICATIONS

Year 1963
Make .... Chrysler
Model... New Yorker Salon
Body style .... two-door hardtop
Base price $5,344
Engine... OHV V-8

Bore x stroke	4.19 x 3.75 inches
CID	413
Compression ratio	10.0:1
Carburetor	Carter AFB 4V
H.P.	340 @ 4600 rpm
Wheelbase	122 inches
Overall length	215.5 inches
Weight	4,290 pounds
Tires	9.00 x 14
OCPG Value	$6,000

The Salon returned in 1964, when it finally showed up in an advertisement. Mechanically, it was no different from other New Yorkers. Chrysler's 413-cubic-inch V-8 with 340 horsepower motivated this 4,300-pound loaded-with-extras car.

CHRYSLER / AMERICA'S NEWEST PACE-SETTERS

Sports spectacular—with a 5-year or 50,000-mile warranty*

It will come as a pleasant surprise to many people that the performance-minded Chrysler 300 comes with America's best and longest new-car warranty—5 years or 50,000 miles.

After all, this Chrysler is built for those people who *really* put a car through the hurdles. But whether the 300 is out winning a rally, pacing the pack at the Indianapolis "500" on May 30th, or enjoying a run down to the beach, it still carries with it that comforting long-term guarantee on all the vital parts that carry its big V-8 power to the rear wheels.

The 300's bucket-seated luxury, alert torsion-bar handling and crisp, custom looks are also there to admire.

Find out for yourself why Chrysler is the *right* choice to set the pace at Indianapolis—and everywhere else performance cars are talked about.

Value spectacular—with a crisp, custom look

No matter how you look at this Chrysler beauty, its value is immediately apparent. Newport's trend-setting styling is still impressing design experts. Its array of standard features never ceases to please prospects and owners alike. Its 5-year/50,000-mile warranty* is the kind of promise only Chrysler engineering can deliver. And, its surprisingly low price is almost impossible to believe.

We still find people who insist we prove it. We're happy to—showing them that by any standard, Chrysler Newport is *the* pace-setter in value.

Stop in and see your Chrysler dealer today. Ask him to *prove* why he's setting the pace in value . . . and why he's setting the pace on deals right now.

*Your authorized Chrysler Dealer's Warranty against defects in material and workmanship on 1963 cars has been expanded to include parts replacement or repair, without charge for required parts or labor, for 5 years or 50,000 miles, whichever comes first, on the engine block, head and internal parts; transmission case and internal parts (excluding manual clutch); torque converter, drive shaft, universal joints (excluding dust covers), rear axle and differential, and rear wheel bearings, provided the vehicle has been serviced at reasonable intervals according to the Chrysler Certified Car Care schedules.

CHRYSLER DIVISION **CHRYSLER** MOTORS CORPORATION

This 1963 Chrysler advertisement highlighted the Indy 500 "Pacesetter" convertible from the Sport 300 series and the low-priced Newport four-door sedan. Five-year 50,000-mile warranty was a big selling point.

The original Salon was a package based on the 1963 Chrysler New Yorker four-door hardtop shown here in a luxurious setting. The Salon included a very generous list of 44 options that we now expect on most luxury cars.

Presenting the crisp, new, custom look of

CHRYSLER '63

Stylists, fashion leaders and designers have hailed the smart, tailored town-car look of Chrysler for '63 as the shape of the future.

And Chrysler's crisp, new, custom look encloses a new world of full-size comfort and quiet, torsion-bar ride control.

Underhood you'll find nothing but V-8 engines that range from the economy of Newport (it uses regular gas and likes it!) to the sports-bred 300

and the spirited luxury of the New Yorker.

Don't look for a small-size Chrysler—we don't make any jr. editions: a fact that protects your pride and your resale value. Tie this in with the good news about prices that start surprisingly low—then read about Chrysler's latest industry first: the 5-year warranty (see adjacent column). Ready? Then take the big, beautiful step ahead —to Chrysler for '63!

Advertisements for 1963 tied Chrysler models to certain lifestyles. The New Yorker four-door hardtop on top is seen in a posh setting. The Sport 300 convertible in the center is seen at a race. The Newport appears in a family type setting.

1949 Crosley: Hotshot prediction

By R. Perry Zavitz

"First by far with a postwar car" was Studebaker's 1947 slogan. First by far with a postwar sports car could have been the Crosley Hotshot's slogan.

Powel Crosley Jr. was the largest maker of radios. He also made refrigerators with shelves in the doors (common today) and owned the Cincinnati Reds. Baseball and ice boxes were not enough to satisfy this go-getter. He dreamed up a tiny car.

The first Crosley was a 12-horsepower two-lung midget. It made its debut, in June 1939, at the New York Worlds Fair. Even though it was a 925-pound weakling, sales amounted to over 2,000 cars before year's end. Many were sold by department stores handling Crosley appliances. By the time car production ended, in early 1942, well over 5,700 Crosleys were on the streets.

The first postwar Crosley was built in May 1946, soon after the first postwar Studebaker. If the Studebaker was revolutionary compared to its predecessor, the Crosley was more so. The body was different, looking more like a real car. It became a real contender in the market.

Crosley's biggest change was not in its body, but in its engine. The two-cylinder, air-cooled, 12-horsepower, Waukesha engine was replaced by a four-cylinder, liquid-cooled engine of its own. Maximum horsepower was 26.5 at a high 5400 rpm. Although displacement was enlarged less that 14 percent, output received a 120 percent boost. Can muscle cars make than kind of boast?

The new engine had overhead valves. There was one surprising feature ... an overhead cam. It was the first American production car since Duesenberg to have one of those.

Unlike any other engine, Crosley's was made of sheet steel. It was developed by the United States Navy. Prototypes included a four the same size as Crosley's and a 250-cubic-inch six that produced 250 horsepower. Crosley went for the small engine for his car. Engineers tinkered and significantly increased power and lowered the maximum rpms.

Engine pieces were stamped of sheet steel, then crimped together. All seams were brazed with pure copper sheets, wires or paste. This required a special 60-foot long furnace operating at over 2,000 degrees. It had three chambers. This copper brazing process gave the engine its name, COBRA. The engine weighed a mere 59 pounds or 150 pounds with accessories.

The Crosley engine developed more power per cubic inch than any other American car at the time. The first car to exceed Crosley's 0.60 horsepower per cubic inch was the 1953 Corvette.

As horsepower was modest, so was torque. Just 32.5 pounds-feet. But, that was outstanding.

Crosley also offered a cast iron engine in 1949, at no difference in price. Specifications were identical to the COBRA engine. However, there were problems with the COBRA. The copper and the steel caused an electrolytic action with pitting that made the thin walls thinner. Some warping was also experienced.

Reluctantly, Crosley went to the cast iron block, Beginning with 1950 models, a CIBA (cast iron block assembly) engine replace the COBRA motor.

Crosley had an economy reputation. Fifty miles per gallon at 35 miles per hour and 35 miles per gallon at 55 miles per hour was claimed. That was optimistic. In reality, at 30 miles per hour, it got about 32.5 miles per gallon and at 45 miles per hour, it got about 30.5 miles per gallon Nevertheless Crosley earned its economy image.

The 1946 two-door sedan was $749. A Chevrolet business coupe cost nearly 47 percent more. Prices rose to $943 in 1952, but by then, Chevy's business coupe cost 62 percent more.

In terms of sales, 1948 was Crosley's high point. There were 25,400 sold that year. One factor that made Crosley popular was a neat little station wagon. Crosley led all car makers in wagon production in 1948.

The next year, Crosley premiered its Hotshot. It was attractive. Many wondered what the Austin-Healey Sprite, of a decade later, would have looked like if there had been no Crosley sports car. Sprites looked like Hotshots with a grille. The two cars were, within an inch, the same size.

The Hotshot had an 85 inch wheelbase; five inches more than the other Crosleys. But, at 136.5 inch overall, it was eight inches shorter. Being so small ... only 3-foot 1-inch to the base of the windshield ... the Hotshot had no doors. There was just a cutout in the side for hopping in or out.

The Hotshot lived up to its name with the CIBA engine. With just 26.5 horsepower, when the light was green could it could get across the intersection before it turned red again? Not to worry! It could go 0 to 60 in 28 seconds and was not much slower than a 1950 MG TD or 1958 Austin-Healey Sprite. With 10.0:1 compression available, the Hotshot had acceleration that could at least equal that. It's been said Hotshot handling was superior to the Sprite's.

Intended for serious competition, the Hotshot's top, headlights, bumpers and windshield could be easily removed. This made the car lighter and more aerodynamic. The Hotshot was courageously entered in several prominent races. The results were noteworthy, if not amazing.

At Sebring, in 1950, Crosley won the Index of Performance. It impressed Briggs Cunningham and he sponsored a special-bodied, souped-up Crosley at LeMans the next year. Its attempt to win the Index of Performance there was sidelined by a defective generator. A Hotshot was run in the Grand de la Suisse race. Fritz Feierabend drove the car at an average speed of 51.2 miles per hour to win the 750-cubic-centimeter and under class.

Another race success was the 1952 Vero Beach 12 Hour Race, in which Briggs Cunningham entered a Crosley-powered Siata. Bob Gegen was one of the drivers who skillfully managed a first-place finish, according to formula. After the race, Eddie Bourgnon, a BSC mechanic, told Gegen that when he went to tear the Crosley engine down, it fell apart in his hands. But, what really counted, was that it lasted the full 12 hours.

Although the name Hotshot is often applied to all the sports cars Crosley made, it is technically incorrect. There was a deluxe version offered. For 1950, it was called Super Hotshot. Then, it was renamed the Super Sports. These featured hinged doors and a folding top. The Super Sports cost about $75 extra.

A total of 2,498 Hotshots, Super Hotshots and Super Sports were built by Crosley in four years. A breakdown by model year shows 752 made in 1949; 742 in 1950; 646 in 1951; and 358 in the abbreviated year of 1952. Neither Nash-Healey or Corvette could produce as many sports cars as Crosley produced.

The Hotshot was years ahead of the competition. A small four-cylinder engine in a tiny domestic sports car has just recently gained acceptance. Can any of these current cars compete as well on the international racing circuits?

The Hotshot was about the size of the smallest oriental cars today, but its weight was about 40 percent less. Crosley's fuel economy was an accurate prediction of today's economy cars. The Hotshot was just too far ahead of its time to benefit from a market not conditioned to accept its outstanding qualities.

SPECIFICATIONS

Year 1949
Make .... Crosley
Model .... Hot Shot
Body style .... two-door roadster
Base price $3,490
Engine .... Inline; four-cylinder
Bore x stroke .... 2.50 x 2.25 inches
CID 44.2
Compression ratio .... 7.8:1 (Manual)
Carburetor .... Tillotson DY-9C 1V
H.P. .... 26.5 @ 5400 rpm
Wheelbase .... 85 inches
Overall length .... 136.5 inches
Weight .... 1,175 pounds
Tires .... 4.50 x 12
OCPG Value $6,000

A 1952 Crosley Super Sports. (Louis Randle)

Lyle Robert's Crosley based 1951 Skorpian.

The 1948 Crosley Hotshot is a perfect clown car for small town holiday parades. (Old Cars photo)

The Hotshot came without doors (and luggage space) in 1950. (Old Cars photo)

DeSoto Adventurer debuted in 1956

By John Lee

Had the DeSoto Adventurer been offered as a convertible, it could have been the pace car for the Indianapolis 500 race on Memorial Day 1956. Instead, the hardtop-only, high-powered and limited-production model paced the Pike's Peak Hill Climb on the Fourth of July. Indy's "Official Pace Car" honors went to one of the Adventurer's sporty companions, a 1956 DeSoto Pacesetter convertible.

The Pacesetter, introduced on January 11, 1956, and the Adventurer (which came on stream some five weeks later on February 18) were both distinguished by their unique paint color combination of a white body with gold side spears and roof. Black-and-gold was optional. In both cases, also finished in gold were the anodized grille and the turbine-style wheelcovers and wheel center medallions. The interior was white, gold and ginger.

The Pacesetter was designed to give DeSoto some extra visibility as a spin-off of the annual "Brickyard Classic " at Indianapolis and help DeSoto dealers sell several hundred replicas of the pace car. The Adventurer had a different purpose; it was DeSoto's answer to the Chrysler 300 letter car.

The 300's debut in early 1955 coincided conveniently with Speed Weeks at Daytona Beach, Florida. There, the hot new number from Chrysler promptly blew away all comers in the Flying Mile and standing-start mile speed trials. The rest of the Chrysler line got into action the following year with the introduction of the DeSoto Adventurer, Dodge D500 and Plymouth Fury early in 1956.

All three of these cars followed the successful Chrysler formula. They featured hopped up engines, beefier suspensions, special paint colors and distinctive interiors in limited-production two-door hardtop body shells.

In DeSoto's case, the hemi-head V-8 got a .05-inch bore increase (from 3.72 inches to 3.78 inches) to boost displacement from the 330 cubic inches in standard form to 341 cubic inches for Adventurers. Dual four-barrel carburetion with smoother manifolding and larger ports aided performance. Also featured were dual exhausts; a higher-than-stock 9.25:1 compression ratio; a higher-performance cam grind; stiffer valve springs; modified pistons; heavy-duty rods; heavy-duty crank pins; and a beefier crankshaft. As a result, the Adventurer's hemi produced a rating of 320 horsepower at 5200 rpm.

Although it was done before Speed Weeks and, thus, not entered in the record books, a DeSoto Adventurer was driven on the beach to a top speed of 137.293 miles per hour. An American passenger car record was set that year by a Chrysler 300B, which was only a shade faster at 139.9 miles per hour.

Aside from the distinctive color treatment, DeSoto's styling was quite similar to 1955. The familiar "toothy" grille was absent for the first time since the end of World War II, giving way to a framed square mesh main element accented by directional signal lamps in vertical bumper guards. Headlight rims had a more pronounced forward peak. The attractive 1955 side trim treatment was revised only to the extent of angling up at the rear to highlight the up-sloped fender peaks and triple taillamp assemblies. Adventurers also had dual antennas accenting the tailfins.

Besides the high-performance engine and suspension, the premium for the Adventurer of $422 over the FireFlite Sportsman hardtop brought the owner power brakes; white sidewall tires; dual outside rear view mirrors; a power front seat; electric windows; a clock; and windshield washers. PowerFlite automatic transmission with a push-button gear selector beside the left-hand windshield pillar was standard on Adventurers, as well as on cars in DeSoto's FireFlite series.

The Adventurer, named for a Ghia-built Chrysler Corporation dream car of 1954, gave DeSoto a true high-performance road machine and a new image for building something beyond the staid, dependable family sedans for which the company had long been recognized.

SPECIFICATIONS

Year	1956
Make	DeSoto
Model	Adventurer
Body style	two-door hardtop
Base price	$3,683
Engine	OHV V-8 (hemi)
Bore x stroke	3.78 x 3.80 inches
CID	341.4
Compression ratio	9.25:1
Carburetor(s)	Two 4V
H.P.	320 @ 5200 rpm
Wheelbase	126 inches
Overall length	220.9 inches
Weight	4,030 pounds
Tires	7.60 x 15
OCPG Value	$15,000

Had the 1956 DeSoto Adventurer been offered as a convertible, it could have been the pace car for the Indianapolis 500 race on Memorial Day. Instead, a special gold and white FireFlite ragtop did the honors. (IMSC photo)

The high-performance Adventurer was DeSoto's answer to the Chrysler 300 letter car. In Adventurers the hemi got a .05-inch bore increase (from 3.72 inches to 3.78 inches) to boost displacement from the 330 cubic inches to 341.

DeSoto's trademark "toothy" grille was absent for the first time since the end of World War II, giving way to a framed square mesh main element. The Adventurer was named for a Ghia-built Chrysler Corporation dream car.

Last DeSoto-built Adventurer

By John A. Gunnell

Most of us think of 1961 as DeSoto's last year and Chrysler buffs know that the last DeSoto was an update of the 1960 Adventurer. However, others will be quick to point out that the last cars to bear the DeSoto nameplate were produced November 30, 1960. Therefore, the title of the story bears further explanation.

The title is based on the less well known fact that the close of 1958 production brought an end to auto manufacturing at the DeSoto factory. The company's 1959 and later cars were actually manufactured in Chrysler plants. Therefore, the last real "DeSoto-built" Adventurer was the 1958 edition.

The fanciest offering to leave the DeSoto factory in 1958 was the Adventurer convertible. It was the company's most expensive product and turned out to be the rarest of the season.

Standard equipment on the Adventurer included a special engine, TorqueFlite automatic transmission, power brakes, dual exhausts, twin outside rear view mirrors, dual rear radio antennas, white sidewall tires, a safety padded dash and distinctive two-tone finish and trim.

The original Adventurer was a limited-production 1956 two-door hardtop featuring a hemi V-8. Only 996 were made. When the 1957 models bowed, the Adventurer had disappeared, but not for long. It returned at midyear in hardtop and convertible models. They had twin four-barrel carburetors and one horsepower per cubic inch was *standard* for the first time ever in any American car.

Adventurer hardtops found 1,650 buyers in 1957. The convertible was even rarer than that, with a mere 300 assembled. Of course, the model was meant to function as an image car, exciting buyers to order other DeSotos.

In 1958, the hemi engine was replaced with MoPar's wedge head V-8, which also carried dual four-barrel carbs. It had a larger cubic inch displacement than the hemi, but the two engines shared the same horsepower rating. A few 1958 Adventurers were equipped with a Bendix electronic fuel-injection system and were later subjected to a recall campaign. Cars with this EFI option were rated at 355 horsepower, before being converted to conventional dual carburetors.

Total production of both body styles in 1958 was a mere 432 cars. Of these, 350 were hardtops. As in the past, both Adventurers were introduced at midyear. This took place during the Chicago Automobile Show, which began on January 4, 1958.

Among the features that distinguish the appearance of Adventurers were gold accented sweep spears with silver trinangle inserts and gold script style model nameplates. There were five bright metal rails decorating the rear deck lid. The special full wheel disks had simulated knock-off hubs.

Options available for 1958 Adventurers included power steeering at $106.30; power windows at $106.30; a Six-Way power seat at $100.65; air conditioning at $492.85; the rare fuel-injection V-8 at $637.20; factory undercoating at $14.30; a rear window defogger for $20.60; and a self-winding steering wheel-mounted clock for $30.15.

SPECIFICATIONS

Year	1958
Make	DeSoto
Model	Adventurer
Body style	two-door convertible
Base price	$4,369
Engine	OHV V-8
Bore x stroke	4.12 x 3.38 inches
CID	361
Compression ratio	10.25:1
Carburetor	(2) Carter 4V
H.P.	345 @ 4000 rpm
Wheelbase	126 inches
Overall length	218.6 inches
Weight	4,180 pounds
Tires	8.50 x 14
OCPG Value	$38,000

This story is based on the fact that DeSoto's 1959 and later cars were actually manufactured in Chrysler and Dodge factories.

The fanciest DeSoto in 1958 was the Adventurer convertible. It was the company's most expensive and rarest car of the season.

1959 DESOTO
IDENTIFICATION FEATURES

ADVENTURER SERIES: "Adventurer" nameplates on front fenders. Gold color sweep inserts and gold finish on grille. Narrow vertical medallion above dip in lower side molding on rear fenders. Metal edging around wheel cut-outs. 2-dr. hardtop roof has simulated Scotch-grained leather finish.

FIREFLITE AND FIREDOME SERIES: Both series have same basic exterior trim, but can be distinguished from one another by series nameplates on front fenders, and by large round medallion above dip in side trim molding on rear fenders of Fireflite models. Silver color sweep inserts optional both series, and side window trim of 4-dr. sedans is bright metal. Fireflite shown above.

FIRESWEEP SERIES: No series nameplates. Silver color sweep insert optional. Painted side window trim on 4-dr. sedan.

The 1959 DeSoto models all had distinct identification features.

1959 brought DeSoto lots of "lasts"

By Bill Siuru

In 1959, automotive wags were predicting that DeSoto, as a marque, was on its last legs. By comparing the 1959 and 1960 DeSoto models, you can see how they came to this conclusion.

About the only good news from DeSoto in 1959 was the fact that it had just produced its two millionth car and the fact that the division was celebrating its 30th anniversary. However, sales were dismal, compared to 1957 and earlier, and the year brought lots of "lasts" for the DeSoto nameplate.

Like all MoPars except the Imperial, this would be the last year for separate body and frame construction. With the adoption of unit body construction came even more commonality between DeSotos and Chryslers. Indeed, many MoPar enthusiasts will tell you that the 26,081 cars built in 1960 and the 3,034 made for 1961 were no more than badge-engineered Chryslers.

Such production totals were a far cry from the 120,000 DeSotos that 1957 wound up with. 1959 became the last year for a separate DeSoto Division. In 1960, the nameplate was merged into the DeSoto-Plymouth Division.

DeSoto would drastically cut the number of models it offered for 1960, but in 1959, there were four series. Firesweep, Firedome, Fireflite and Adventurer were the series names, in order of prestige and price. Only the Fireflite and Adventurer would survive into 1960. By 1961 all of the models were just plain DeSotos.

Also to disappear in the last two years were convertibles and station wagons. Only the two- and four-door hardtops and four-door sedan would survive through 1960. Only the two hardtops remained for 1960. However, for 1959, ragtops were available in all series. Station wagons were also available in Firesweep and Fireflite versions, with either six- or nine-passenger seating. 1959 was also the last year that hardtops were given the Sportsman designation.

The 1959 DeSoto Firesweep Sportsman two-door hardtop was the second lowest priced DeSoto model. At a factory price of $2,967, it was only about $60 more than the Firesweep four-door sedan.

All Firesweeps came with the 361-cubic-inch "wedge head" V-8 that produced 295 horsepower at 4600 rpm. This was accomplished with a 10.0:1 compression ratio and a two-barrel carburetor. Other 1959 DeSotos used the larger-bore 383-cubic-inch V-8 version of the engine, which produced 305 to 350 horsepower, depending on series.

While optional on Firesweeps, most cars were still equipped with TorqueFlite or PowerFlite automatic transmissions. They came complete with push-button controls on the dashboard.

The Firesweeps were the only 1959 DeSotos to ride a 122-inch wheelbase chassis derived from the Dodge. The rest of the DeSotos used the same 126-inch wheelbase chassis that Chryslers, except Windsors, employed. (The Windsor also had a Dodge-type chassis).

Even at that, Firesweeps were just a few inches shorter than other DeSotos. Of course, there was no shortage when it came to tailfins. They were large and still in vogue. Chrome and color side moldings were also optional on Firesweeps.

SPECIFICATIONS

Year	1959
Make	DeSoto
Model	Firesweep
Body style	Sportsman two-door hardtop
Base price	$2,967
Engine	OHV V-8
Bore x stroke	4.12 x 3.38 inches
CID	361
Compression ratio	10.0:1
Carburetor	Carter BBD 2V
H.P.	295 @ 4600 rpm
Wheelbase	122 inches
Overall length	217.1 inches
Weight	3,625 pounds
Tires	8.00 x 14
OCPG Value	$9,500

In 1959, automotive wags were predicting that DeSoto, as a marque, was on its last legs.

The 1959 DeSoto Firesweep Sportsman two-door hardtop was the second lowest priced DeSoto model. There was enough room in the rear to sit with your hat on.

The last year for a separate DeSoto Division was 1959. In 1960, the nameplate was merged into the DeSoto-Plymouth Division.

All Firesweeps came with the 361-cubic-inch "wedge head" V-8 that produced 295 horsepower at 4600 rpm.

Dodge dynamite: the D-500

By Robert C. Ackerson

It's common fare for most automotive histories to focus on either the 1955 or 1957 model years when they examine the postwar cars from Chrysler Corporation. The logic behind this perspective is based upon the significance of those two years in Chrysler history.

In 1955, Chrysler broke away from the bondage of its initially conservative postwar styling strategy with a series of attractively designed cars. Just two years later, Chrysler did it all over again with its "Forward Look" of 1957.

Sandwiched between these two years were the 1956 models. They appear, it seems, like little more than footnotes to Chrysler's drive for styling superiority. In terms of production, the 1956 MoPars pale in comparison to the 1955 and 1957 models.

The relatively weak sales of the 1956 models represented more a consequence of over-selling the 1955 models, than a deficiency in the 1956 Chrysler products. The over-selling left the market drained of new car buyers. 1956 may not have been a vintage year for automotive design, but it was a year when Detroit, building on the strength of 1955, made dramatic strides in the area of high-performance.

Although Chrysler Corporation never followed the lead of General Motors and Ford in marketing two-seat sports cars, in 1956 it did expand its own style of high-performance cars ... the Chrysler 300 ... by introducing similar models from Plymouth, Dodge and DeSoto.

Of this trio, the Dodge D-500 was, by far, the most understated. Unlike the Plymouth Fury and DeSoto Adventurer, its exterior identification was limited to small emblems (checkered flags linked by "500" numerals) mounted on the hood and rear deck. Actually, the D-500 wasn't even a special model. Instead, it was an engine-suspension package.

Other Dodge V-8 engines for 1956 reverted to the less costly polyspherical head design with single rocker shafts. However, the D-500 engine continued to use the hemispherical combustion chambers and double rocker shafts found in the 1955 Super Red Ram V-8. With a 3.63 inch bore and 3.89 inch stroke, the D-500 engine displaced 315 cubic inches. Its compression ratio was a healthy 9.75:1 and a four-barrel Carter WCFB carburetor was installed. The intake valve diameter was 1.875 inches, while the exhaust valves measured 1.53 inches.

Dodge rated this engine at 260 horsepower at 4800 rpm and 300 pounds-feet of torque at 3000 rpm. Contributing to this output was a dual exhaust system with 1.75-inch diameter pipes.

The standard D-500 transmission was the not-too-exciting Chrysler heavy-duty three-speed manual, whose advanced age was revealed by its poorly-spaced ratios of 2.5:1 (first), 1.68:1 (second) and 1:1 (third). The optional PowerFlite automatic transmission was also less than a perfect performance match up with the 260-horsepower V-8, since it was just a two-speed unit. It did, however, have higher shift points than the PowerFlites used in other Dodge models.

While it could not equal the outstanding roadability of the 1957 Chrysler products with their front torsion bar suspensions, the 1956 D-500 suspension provided excellent handling and stability. There was nothing unusual in its design of front coil springs and unequal length arms along with a solid rear axle and semi-elliptic leaf springs. However, this run-of-the-mill design was given a much higher level of competence, due to the use of stronger-than-stock steering arms, knuckles and axle shafts; a 0.8125-inch diameter front anti-roll bar; plus, super-stiff springs and heavy-duty Oriflow shock absorbers normally used on Dodge police cars.

The front springs were constructed of 6.75-inch diameter steel and were shorter than normal. The six rear leaf springs were flatter than normal, too. These revisions gave the D-500 a height approximately two inches lower than other 1956 Dodge models.

Also in tune with the D-500's high-speed capability was the installation of a first-class braking system. It consisted of two-leading-shoe front brakes and 12 x 2.5-inch drums, that were normally used on heavier Chrysler models, at both the front and rear. These brakes provided a total effective area of 252 square inches, which was far more than the 173.5 square inch area of other 1956 Dodge models.

Hot Rod magazine (May 1956) noted that "this means a 45 percent increase of lining area or 15.5 pounds of car weight per square inch of lining area, a figure without equal in American cars and one that approaches some sports and racing types."

Sports Cars Illustrated (August 1956) noted that "Chrysler deserves some sort of citation for giving their hot cars stopping power proportional to the going power."

As far as *Hot Rod* was concerned, the D-500 was without peer (among the cars it had tested to that date) in regard to its braking quality and acceleration ability. Its 0 to 60 time was 8.8 seconds with the 0 to 80 run requiring 15.6 seconds. Its best speed in the standing-start quarter-mile was 83.6 miles per hour. At that time a D-500 equipped two-door Coronet with manual transmission, optional high-lift camshaft and a 4.56:1 axle ratio (3.73:1 was standard with either manual or automatic transmission) held its class record at the San Fernando dragstrip, in California, with a speed of 84.6 miles per hour.

Since just about every name even remotely related to high-performance activities was eventually used by American manufacturers to identify their super stock models, it wasn't surprising that Dodge used the D-500 label for its entry into the world of factory hot rods. In 1954, use of a special Dodge Royal convertible to pace the Indianapolis 500 race had led to the production of 701 replica models with such items as Kelsey-Hayes chrome wire wheels, continental tire mount, special door trim and checkered flag "500" emblems.

In 1956, the D-500 label was also linked to the NASCAR regulation calling for a miminum production run of 500 units for a car to be eligible for NASCAR stock car competition. While the D-500's availability satisifed this requirement, an even more Potent D-500-1 package was provided for NASCAR racing. It's primary components consisted of a dual carburetor intake manifold carrying two Carter WCFB four-barrel carburetors and a higher-lift cam (available separately as part number 173255) with 60-degrees of overlap and 280-degrees intake and 270 degrees exhaust duration. Its horsepower peaked at 295.

Yes Virginia ... there are some real D-500s left today!

At the 1956 NASCAR Speed Weeks, a D-500-1 set a Flying Mile class record of 130.577 miles per hour. An earlier one-way run resulted in a maximum speed of 133 miles per hour. In the one-mile standing-start acceleration runs, the Dodge was the fastest passenger car entered. Its 81.7 miles per hour speed compared favorably with an 88.8 miles per hour run, made by Chuck Daigh in a Thunderbird running without its standard bumpers, windshield and spare tire mount.

When the 1950s began, Dodge was still a conservative automobile powered by an antiquated L-head six-cylinder engine with models carrying such names as Meadowbrook and Coronet. With the arrival of the Red Ram V-8 in 1953, Dodge began its transformation into an automobile with a more contemporary image. The introduction of the D-500, in 1956, clearly indicated that the changeover was complete.

SPECIFICATIONS

Year 1956
Make .... Dodge
Model .... Coronet D500
Body style .... two-door sedan
Base price $2,267
Engine .... OHV V-8
Bore x stroke .... 3.63 x 3.89 inches
CID 315
Compression ratio 9.75:1
Carburetor .... Carter WCFB 4V
H.P. .... 260 @ 4800 rpm
Wheelbase .... 120 inches
Overall length .... 212.0 inches
Weight .... 3,380 pounds
Tires .... 7.10 x 15
OCPG Value $7,800

This is your reward for the great Dodge advance—the daring new, dramatic new '56 Dodge

The Magic Touch of Tomorrow!

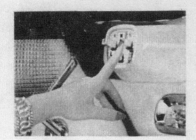

The *look* of success! The *feel* of success! The *power* of success!

They come to you in a dramatically beautiful, dynamically powered new Dodge that introduces the ease and safety of push-button driving —the Magic Touch of Tomorrow! It is a truly great value.

New '56 DODGE
→ VALUE LEADER OF THE FORWARD LOOK

"The Magic Touch of Tomorrow" was the theme used to advertise Dodge's push-button automatic transmission in 1956. Dodge was called "Value Leader of the Forward-Look" and the D-500 option made it a muscle car.

The Royal Lancer hardtop seen in this 1956 advertisement was for the young buyer who wanted to look his or her best when cruising. Those who wanted to cruise fast added the D-500 package.

A Dodge D-500 sedan from the Kiekhaefer Mercury Outboards stock car racing team takes the checkered flag on the sands of Daytona Beach in 1956. Note the D-500 crossed flags emblem on hood. (Mercury Marine photo via Ray Doern)

MoPar's early muscle: 1962 Dodge

By Dave Duricy, Jr.

Chrysler has sometimes suffered from a problem: when it offered an outstanding car, the public went in another direction in droves. Even with the advantage of hindsight, collectors seldom have reversed that trend. One such car is the focus of our attention now.

During the early 1960s, Dodge was going through one of its roughest times regarding identity. Through 1959, Dodge was the next step up from Plymouth in the Chrysler pecking order. But a curve was thrown at the public when the redesigned 1960 Dodge came in two sizes.

One was the traditional larger-than-Plymouth line of Matador and Polara models. A new based-on-Plymouth series was also introduced. Given the name Dart, it included Seneca, Pioneer, and Phoenix models. (The Dart is usually thought of as Dodge's compact line, but the first ones were not compacts. Compact Darts first appeared in model-year 1963.)

While styling was new for 1960, there was nothing in the appearance of either of the two car-lines that said "Dodge." Nor was there any close relationship between the two to tell potential customers that both of these cars were the new Dodges.

Though totally new grilles appeared on the 1962 models, a needed revision was made; both sizes of Dodges now used the same style grille. The upper series was slightly fancier, which was appropriate for a larger and more expensive car.

However, the 1962 front end styling bore no relationship to previous Dodges. Therefore, would be buyers had to read the name on the 1962s to know that these were indeed Dodges. (Styling and other changes for 1963 would add to Dodge's identity confusion. The large models continued the 1962 front end styling, but the name was changed to Custom 880.)

The 1962 Polara was dropped down to become the top line for the smaller Dodges. It was called Polara 500, to be more specific. While it had the size and styling of the Dart, it was not considered a Dart.

The smaller 1962 Dodges were redesigned and down-sized. The wheelbase was shortened two inches and overall length shrunk almost 7-1/2 inches. Because of the recent introduction of compact cars, and the continued enlargement of full-size models, a new and needed mid-size class was emerging. The Dart spearheaded Dodge's entrance into this size.

The Senaca, Pioneer and Phoenix names were dropped from the 1962 Dart series. Instead, the base version was left nameless, the middle range was called the Dart 330 and the higher-priced edition was the Dart 440. Those digital titles did not seem to have any specific numeric value. But there was a special Dart model that was known by a number that had a very specific meaning. That was the Dart "413" or "Ramcharger 413." This designation referred to the displacement of an optional V-8 high-performance engine that powered some of the mid-size Dodges.

With the 1960 introduction of the Dart, Dodge offered a choice of engines for it. The base engine was the new-in-1960 "slant" six. Optional were the 318-cubic-inch V-8 and the 361-cubic-inch V-8. In addition, the "small" 383-cubic-inch "big-block" with 325 or 330 horsepower was available in the 1960 Dart Phoenix. For the following year, Chrysler's 413-cubic-inch V-8 (with 350 or 375 horsepower) could be ordered in a Dart.

This same engine was offered in the 1962 Dart. However, in the 1962 model, its availability gained added historic significance. Since the Dart was a mid-size model, this was technically the first muscle car, in the sense that this term was used in the 1960s.

The 1964 Pontiac GTO is usually credited with starting the muscle car era. What John Z. DeLorean did for Pontiac, to create the GTO, was pull the 389-cubic-inch V-8 from the full-size Pontiac and put it into the mid-size Tempest. This made the Tempest intermediate a "factory hot rod" or muscle car.

Perhaps DeLorean got some of his inspiration from the things that Dodge did two years earlier. He was smart enough and keen enough on performance that he surely knew what Dodge was doing back in 1962. Let's compare the 1962 Dodge Dart 413 Ramcharger to the 1964 Pontiac GTO.

The Dart had a 116-inch wheelbase and was 202 inches long. That was an inch longer wheelbase, but an inch shorter overall length than the GTO. The Dart tipped the scales at 3,890 pounds, which was 90 pounds heavier than the GTO. They were virtually the same size and almost the same weight.

Obviously, the Dodge's 413-cubic-inch displacement was larger than the GTO's 389. Not surprisingly, output was also more. Rated horsepower was 385 for the Dart, compared to 348 in the GTO. The difference in horsepower was greater than the difference in displacement.

Some of that difference may be explained by compression. The Dart had 11.0:1 compression, while the GTO was slightly less at 10.75:1. However, the torque was proportional; 455 pounds-feet for the Dart and 428 pounds-feet for the GTO.

Of course, the main point is performance. Although various magazines tested one or the other of these cars, *Car Life* tested both. Its tests are probably the best comparison available. The cars varied mainly in the type of transmission used. Dodge offered a four-speed transmission, but the Dart 413 tested had a three-speed. The GTO had a four-speed.

Zero to 60 miles per hour times were 7.4 seconds for the Dart and 6.6 seconds for the GTO. In the standing-start quarter-mile, the Dart need 15.1 seconds to cover the distance at a terminal speed of 92 miles per hour. The GTO did it in 14.8 seconds at 99 miles per hour. Top speeds were 107 miles per hour (Dart) and 135 miles per hour (GTO). The Dart gave 10-13 miles per gallon, versus 12-15 for the GTO.

One of the main differences between the performance of the two was the type of transmission. Another factor was a variation in final drive ratios. However, it must be admitted that the Dodge 413 was an excellent performer. It was a great move in the direction of muscle cars and a model for Pontiac to improve upon.

Certainly, it was the GTO that popularized muscle cars. There were 32,450 sold in the model's first year. That amounted to one 1964 GTO selling for every five 1962 Dodges (all series and engines). However, the number of those Dodges equipped with the 413-cubic-inch engine was most likely quite limited. In the case of the GTO, the big engine was standard equipment.

SPECIFICATIONS

Year	1962
Make	Dodge
Model	Dart 440
Body style	two-door hardtop
Base price	$2,731
Base V-8 engine	OHV V-8
Bore x stroke	3.91 x 3.31 inches
CID	318
Compression ratio	9.0:1
Carburetor	Carter BBD 2V
H.P.	230 @ 4400 rpm
Wheelbase	116 inches
Overall length	202 inches
Weight	3,185 pounds
Tires	6.50 x 14
OCPG Value	$6,000

SPECIFICATIONS (Ramcharger 413)

Year	1962
Make	Dodge
Model	Dart "413 Ramcharger"
Body style	two-door hardtop
App. Base price	$3,500
Base V-8 engine	"Max-Wedge" OHV V-8
Bore x stroke	4.19 x 3.75 inches
CID	413
Compression ratio	11.0:1
Carburetor	Two Carter 4V
H.P.	410 @ 5400 rpm
Wheelbase	116 inches
Overall length	202 inches
App. Weight	3,300 pounds (steel)
Tires	8.00 x 14
OCPG Value	$26,000

The 1962 front end bore no relationship to previous Dodges. Buyers had to read the name on the 1962s to know that these were indeed Dodges.

The small 1962 Dodges were redesigned. The Lancer (rear view) was a compact. The Dart 440 (fore-ground) was in the middle. Its wheelbase was shortened two inches from 1963 Dodges and overall length shrunk almost 7-1/2 inches.

The down-sized 1962 Dodge Dart had a 116-inch wheelbase and measured 202 inches long overall. Though a six was standard in this Dart 440 four-door hardtop, the 413-cubic-inch big-block V-8 was optional in many models.

From the rear, the 1962 Dodge Dart 440 four-door hardtop had the popular Lancer look applied to a larger size body. Even the simulated louver rear quarter trim and twin-stacked round rear lamps were similar to Lancer's.

Just because Dodge's debonair two-door looked like a muscle-bound machine and had the power to boot, that didn't mean that it forsook passenger comfort or practicality. This view shows the headlamps hidden away.

Charger: Dodge's first-class fastback

By Dave Duricy, Jr.

In 1964, Plymouth introduced its "glassy" fastback, the Barracuda. Ford brought out its nifty fastback version of the Mustang for 1965 and even AMC tried the body style in the form of the 1965 Marlin. These models helped spark a craze that had last been see in the early 1950s ... the fastback body style.

The 1966 Dodge Charger appeared at the height of this craze and wore what could probably be considered the last word in fastback styling. It was based on the 117-inch wheelbase Dodge Coronet and shared many mechanical, structural and body parts. However, Dodge made certain that the Charger compromised nothing when it came to performance or image.

The highlight of the Charger package was, without a doubt, its sweeping silhouette. It was dominated by a sloping roof line which began at the windshield header and ran the full length of the car in a gentle, down-turned curve which terminated just above the full-width taillight.

This roof styling resulted in long, slim roof pillars and a vast expanse of side glass. Combined with the pillarless body style, this provided great vistas of the outside world. The effect the roof gave was one of speed and motion. Even standing still, the Charger seemed ready to lunge forth in a great burst of speed.

Dodge made certain that was possible. Aside from the Charger's standard 318-cubic-inch, 230-horsepower V-8 power plant, 1966 buyers had the option of choosing from a group of engines. There was a 361-cubic-inch, 265-horsepower V-8; two versions of MoPar's 383-cubic-inch V-8 (one rated at 270 horsepower and the other at 325 horsepower); as well as the 426-cubic-inch hemi V-8 with its rip-snorting 425 horsepower.

Just because Dodge's debonair two-door looked like a muscle-bound machine and had the power to boot, that didn't mean that it forsook passenger comfort or practicality. Charger's interior boasted four-place bucket seats covered in high-quality vinyl.

A center armrest and ashtray were provided for rear passengers, and the cabin was fully carpeted. The left outside mirror was remote-controlled. The Charger interior was equipped with a padded dash and the expected bright interior accents and courtesy lights.

The most intriguing aspect of the Charger's interior was its cargo-carrying ability. The rear seat backs could be folded down to provide a vast flat area that ran from the trunk to the back of the front seats. Dodge described the space created by this ingenious interior plan as being "wagon like." The fastback roof was not only stylish, but functional as well.

A Charger could be customized in any number of ways. Its 1967 brochure stated that the Charger was available in 18 acrylic colors, plus a special-buffed silver paint job at extra cost. The same brochure describes interiors available in red, blue, white, black, copper and combinations of black and gold. Also mentioned are three different style wheels including chrome steel Road Wheels (mags).

Black or white vinyl top coverings, electric windows, radios, air conditioning, a four-speed manual transmission, front disc brakes, plus a dizzying array of rear axle ratios, suspension options and tires were also available.

The Charger was a lot of automobile, but not too much for Dodge, at the 1966 base price of $3,122. After a 1966 run of 37,344 examples and 15,788 of the 1967 models, Dodge decided to let the Charger continue to lead the "Dodge Rebellion" into 1968. A new body with a less over-powering fastback and curvaceous styling brought new meaning to the word "wow."

The 1966-1967 Charger has earned a place in automotive history because of its daring profile and sleek, good looks. It was an elegant specialty coupe that was too big to be called a "pony car" and too "hot" to be labeled a "personal luxury" automobile. The first Chargers are probably simply (and best) described as being "first-class fastbacks."

SPECIFICATIONS

Year 1966
Make .... Dodge
Model .... Charger
Body style .... two-door hardtop
Base price $3,122
Engine .... OHV V-8
Bore x stroke .... 3.91 x 3.31 inches
CID 318
Compression ratio 9.0:1
Carburetor .... Stromberg 2V
H.P. .... 230 @ 4400 rpm
Wheelbase .... 117 inches
Overall length .... 203 inches
Weight .... 3,499 pounds
Tires .... 7.75 x 14
OCPG Value $14,000

SPECIFICATIONS

Year 1967
Make .... Dodge
Model .... Charger
Body style .... two-door hardtop
Base price $3,128
Engine .... OHV V-8
Bore x stroke .... 3.91 x 3.31 inches
CID 318
Compression ratio 9.0:1
Carburetor .... Stromberg 2V
H.P. .... 230 @ 4400 rpm
Wheelbase .... 117 inches
Overall length .... 203 inches
Weight .... 3,480 pounds
Tires .... 7.75 x 14
OCPG Value $13,000

The 1966 Charger earned a place in history with its daring profile and good looks. It was an elegant specialty coupe, too big to be a "pony car" and too "hot" to be labeled a "personal luxury" automobile. (Chrysler Historical)

The new-for-1966 Dodge Charger wore what could probably be considered the last word in fastback styling. The highlight of the Charger package was, without a doubt, its sweeping silhouette. (Chrysler Historical)

Late-model convertibles seem to be over-priced and impractical road machines, with the possible exception of the Ford Mustang. The author would rather have his old 1939 Ford convertible and keep the change.

1939 Ford ragtop is oldest, yet latest thing

By Ned Comstock

Forbes magazine once presented an interesting update on the convertible, the death of which seemed to have been greatly exaggerated, not too long ago. In the late-1970s, the ragtop was considered doomed, and thought to be virtually extinct. By 1987, automakers were offering at least 20 convertibles and their sales were projected at over 100,000 units, including imports. That was up from nearly zero a decade earlier.

Forbes explanation of the convertible's near escape from oblivion, two decades ago, was probably right. According to the magazine, after 1965 (the high point for convertible sales) there came an emphasis on air conditioning, higher speeds and driver safety, all of which pushed buyers toward steel tops.

However, there have always been buyers who like to feel the outdoors when they're driving. That demand is still there and as strong as ever. That's not the problem. The problem comes with the new cars themselves.

Late-model convertibles seem to be the most over-priced and impractical road machines you can imagine, with the possible exception of the Ford Mustang. I'd rather have my 1939 Ford convertible and keep the change.

First, take the price, because everybody but a few Arabian oil sheiks does. In 1939, according to *The Standard Catalog of American Cars* (Krause Publications), Ford's convertible sold new for $5 more than the $790 Fordor sedan. This proves that there is nothing inherently expensive about convertible manufacturing. Yet, the estimated cost of the Cadillac Allante convertible, at $50,000, was twice the cost of a standard Cadillac sedan. That was a $25,000 premium, for which the convertible buyer got nothing, but the hype that goes with the sale.

It's hard to bring 1939 prices into modern times, but the price of gold is the best constant we have. Using a factor of 10, which is how much gold has risen, a Ford convertible was priced, in 1939, at $7,950 in today's dollars. Contrast that with the 1987 Ford Mustang ragtop at $16,500. Clearly, the first strike against the late-model convertible buyer is price. Next comes practicality.

Albert Einstein said everything should be as simple as possible, but not simpler. He was right. I thought of him on a long, below-zero-degrees, ride home in my modern convertible one New Year's Eve. The windows were frozen in the down position, where some prankster had put them.

In recognition of the Einstein dictum, there was nothing on the 1939 Ford convertible that didn't contribute to the car's going or the driver's real need. The manual transmission, for example, was controlled by a stick shift that sprouted out of the floor. It gave the driver the best possible control of his engine and freed him from transmission trouble forever. As if to prove the advantage of manual transmission, years after Hydra-Matic had practically taken over the industry, "four-on-the-floor" became a popular item of extra-cost equipment. It was eagerly sought by drivers who were seeking superior performance.

Engineering, the application of scientific principles, is another reason why the 1939 Ford convertible seems superior now. Once the abominable mechanical brakes of previous Fords had been replaced with 1939 hydraulics, the car design was modern, just as it would be today.

The 85 horsepower engine was plenty to keep up with highway traffic then and it would be now. Later experience with two Porsches found their higher speed no advantage and their cramped passenger quarters uncomfortable, compared to my 1939 Ford. Indeed, if the 1987 Porsche 911 cabriolet (estimated price $47,000) could double the Ford's speed, it could not be said it was going anywhere. And the Ford V-8 was an easy running, long-lived engine that packed in place neatly and required little maintenance. Except for transverse-mounted engines and front-wheel-drive, late-model cars have little that's new and really needed.

Another factor in 1939 Ford success was style. The car looked so good you could almost bet that Edsel Ford had something to do with it and, sure enough, he was Ford president that year.

My 1939 Ford convertible was bottle green, set off by the restrained use of chrome. The top was taut and well-tailored. The hood was short, with a pronounced "V" and the passenger space set forward, giving the car an eager, thrusting look. It looked like an automobile, too, and not like a guided missile or a rocket ship.

It's natural to wonder that, if the 1939 Ford convertible was such a great car, why was it discontinued nearly 50 years ago? The answer comes easily. The American auto industry, with its over-priced, complicated, road rockets has been on the wrong track for 50 years. Proof of this comes easy. *Old Cars Price Guide* list the 1939 Ford convertible as selling at $28,000 today, if you can find one. That's what people really think of a simple, practical good-looking car.

SPECIFICATIONS

Year 1939
Make .... Ford
Model .... Deluxe
Body style .... Convertible
Base price $790
Engine .... L-head V-8
Bore x stroke .... 3-1/16 x 3-3/4 inches
CID 221
Compression ratio 6.15:1
Carburetor .... Stromberg 2V
H.P. .... 60 @ 3500 rpm
Wheelbase .... 112 inches
Overall length .... 179.5 inches
Weight .... 2,840 pounds
Tires .... 6.00 x 16
OCPG Value $28,000

There was nothing on the 1939 Ford convertible that didn't contribute to the car's going or the driver's real need.

The hood was short, with a pronounced "V" and the passenger space set forward, giving the car an eager, thrusting look. It looked like an automobile, too, and not like a guided missile or a rocket ship.

All Ford Deluxes used a distinctive grille. It had horizontal bars in the center portion, flanked by side grilles with smaller horizontal segments. Cars equipped with radio had antenna on center of windshield header.

1940 Ford Deluxe was handsomely new

By John A. Gunnell

Actually introduced on October 6, 1939, the "forty Ford" was far from innovative in a mechanical sense, but excelled when it came to good looks and classic design.

Ford took a basically tried-and-true package, added a few improvements, and dressed-up the results with appearance features boasting timeless appeal to automotive enthusiasts.

As George H. Dammann notes in his *Illustrated History of Ford* (Crestline Publishing Company) the 1940 Fords were "proving tremendously popular, first to the new-car buyers and, later on, to a wide variety of used car customers ... especially hot rodders, stock car racers and, primarily in the southern mountain regions, to moonshine runners." Today, they are equally appealing to collectors.

Two car-lines were marketed for 1940. The low-priced Ford passenger car line was the 022A or standard series. These cars looked somewhat like the deluxe 1939 models. The higher-priced cars were designated as the 01A or deluxe series. They featured a body with handsome new styling rendered by designer Eugene "Bob" Gregorie.

Deluxe equipment added 60 pounds of extra trim items and equipment, including chrome headlamp doors, red-trimmed Ford Deluxe hubcaps, twin taillamps (except wagons), a stem-wind clock, bright metal wheel trim bands, a left front door armrest (except wagons) and dual sun visors. Inside the Deluxe had a two-tone dash and matching steering wheel, instead of the standard Briarwood Brown dash and wheel. The deluxe instrument panel face was also smaller.

All Deluxes featured Ford's larger 85-horsepower engine. A distinctive grille was seen at their front end. It had horizontal bars in the center portion, flanked by side grilles with smaller horizontal segments. (The standard had a 60-horsepower V-8 and 1939-style grille).

The Ford Deluxe shared the 112-inch wheelbase with standard models, as well as a 123.13-inch spring base and 190.86-inch bumper-to-bumper length. Overall width was 69.69 inches. The Fordor Sedan weighed just under 3,000 pounds.

Both of Ford's flathead V-8s had three main bearings, Chandler Grove carburetors and Ford ignition systems. The more powerful motor was larger of course. It had higher and smoother compression and torque curves and a steeper, higher horsepower curve. Hydraulic brakes were featured.

Both lines offered a three-passenger coupe, a Tudor Sedan, a Fordor Sedan and a Station Wagon. In the Deluxe line only, buyers could also obtain a five-passenger Business Coupe and a Convertible Coupe. The ragtop no longer had a rumbleseat and the Ford Deluxe series no longer included a Convertible Sedan.

Despite the lack of drastic technical changes, Ford made improvements that were rather heavily promoted as advances for 1940. They included a controlled ventilation system with pivoting wind wings, a fixed windshield, cowl-mounted windshield wipers, a two-spoke steering wheel, a column-mounted gearshifter, new wheel rims of curved disc design, smaller hubcaps, front suspension upgrades and a nine-inch centrifugal clutch.

Pricing for Ford Deluxes began at $722 and reached to $947. The highly collectible Convertible Coupe sold for $849 new. Today, one of these ragtops in perfect condition is very hard to find for sale and super-expensive when offered.

During the 1940 model-year, Ford cranked-out 575,085 total cars. In the Deluxe series, the production total included 23,704 Convertible Coupes; 27,919 three-passenger coupes; 171,368 Tudor Sedans; 91,756 Fordor sedans; 8,730 Station Wagons; and 20,183 Business Coupes.

The cars are very enjoyable to own and operate. When maintained properly, the 85-horsepower flathead V-8 is fairly quiet and smooth-running. The interior is comfortable and attractive, since the company was trying to give all of its deluxe models a more Lincoln-like feeling back then. On the exterior, Ford Deluxe buyers could get the three standard colors of Black, Lyon Blue or Cloudmist Gray, plus exclusive deluxe shades of Folkestone Gray, Yosemite Green and Mandarin Maroon.

SPECIFICATIONS

Year 1940
Make .... Ford
Model .... 01A Deluxe
Body style .... Convertible Coupe
Base price $850
Engine .... L-head V-8
Bore x stroke .... 3-1/16 x 3-3/4 inches
CID 221
Compression ratio 6.15:1
Carburetor .... 2V downdraft
H.P. .... 85 @ 3800 rpm
Wheelbase .... 112 inches
Overall length .... 188.25 inches
Weight .... 2,956 pounds
Tires .... 6.00 x 16
OCPG Value $29,000

Ford Deluxes shared the 112-inch wheelbase with standard models, as well as a 123.13-inch spring base and 190.86-inch bumper-to-bumper length. The Ford Deluxe was a different car than the Standard, which had 1939-like styling.

There was a lot of old-fashioned "horse sense" behind the 1940 convertible. Ford took a basically tried-and-true package, added improvements, and dressed-up the results with timeless design appeal. (Ford Archives)

The higher-priced cars featured a body with handsome new styling rendered by designer Bob Gregorie. The car looked even better in person than it did in the design renderings. (Ford Archives)

Ford had a high-style sedan in 1950

By John Lee

Earl was a young preacher we knew in the 1950s. My folks looked forward towards Earl's regular visits, because he was such a pleasant and inspiring young man. I enjoyed having him come, too, even though the grown-up talk didn't mean much to me. I was always anxious to see Earl's car!

He was actually still a seminary student who drove out, on Friday, to serve the church in a little town seven miles down the road. He then went back to his campus on Sunday evening. Since my father was also a minister, Earl dropped by on his way to or from his parish, to visit and seek a word of encouragement or borrow a book.

Earl wasn't a typical, impoverished theological student, however. He wasn't paying tuition, buying gasoline and feeding himself and a wife on the salary of a weekend pastor. I guess it was his family who saw to it that he always had adequate transportation. The weekly round trip was, after all, 350 miles.

So, Earl always had a new car. At least, he did have one during the two years I remember his regular visits. They were Fords, and not stripped economy models either. He drove a 1950 Custom Deluxe Fordor sedan which listed at $1,637 before options were added. Only the specially trimmed Crestliner, the convertible and the wood-bodied station wagon had higher price tags.

His car had the 100 horsepower L-head V-8 engine with a Holley two-barrel. An L-head six was available, but anyone going first class in a Ford would get the engine that had given Ford its performance reputation since its introduction in 1932. There was no way of telling, from the outside, whether Earl's car had optional overdrive with its three-speed transmission. This option reduced engine revolutions by about 30 percent. Probably, the car had it, since most of his traveling was over-the-road miles. You couldn't get an automatic in a Ford yet.

The young preacher's sedan was finished in medium green. What was really impressive was the full load of genuine Ford accessories that gave it a customized look. It had a radio, of course (we didn't even have one in our 1946 Dodge), whitewall tires and wheel trim rings. Fog lights were mounted in front and a back-up light in the rear. It may even have had a sun visor, outside rearview mirror and a spotlight ... I don't remember. But, I do remember the neat, three-ribbed rear fender skirts. These had been introduced to dress-up the high-style Crestliner and were offered optionally on other models.

When he traded and got a new Ford in 1951, Earl got the same model and virtually the same equipment, but in a rich, metallic brown color.

The 1950 Ford was one of 247,181 Custom Deluxe Fordors produced that year. It was the third most popular choice, falling behind the Custom Deluxe Tudor sedan (398,060 units) and the Deluxe Tudor sedan (275,360 units).

All us kids who kept up on such things could tell a 1950 Ford from a 1949 easily enough, even though the grown-ups thought they looked the same. From the front there was a new Ford crest on the hood in place of the F-O-R-D letters curving over the center of the grille and the main grille bar curved around the corners of the fenders along with new parking light housings.

On the side view, the door handles were of push-button design, while the 1949s had been pull-type. The front piece of the chrome spear bearing the "Custom" designation was redesigned and the gas filler was hidden inside a trap door, instead of being exposed. Around the back, a new, painted, deck lid handle and license housing with a Ford crest above it replaced the propeller-looking chrome handle of 1949.

In spite of the design being a year old, the Ford Custom Deluxe was about as stylish as you could get in 1950. And Earl's fully-accessorized Fordor was the epitome of style.

SPECIFICATIONS

Year	1950
Make	Ford
Model	Custom Deluxe
Body style	four-door sedan
Base price	$1,637
Engine	L-head V-8
Bore x stroke	3.19 x 3.75 inches
CID	239
Compression ratio	6.8:1 (Manual)
Carburetor	Holley 2V
H.P.	100 @ 3600 rpm
Wheelbase	114 inches
Overall length	196.8 inches
Weight	3,093 pounds
Tires	6.70 x 15
OCPG Value	$7,400

Earl drove a 1950 Custom Deluxe Fordor sedan like this one, except that his had fender skirts and lots of accessories. With the 100-horsepower L-head V-8 engine and Holley two-barrel it listed for $1,637. (Ford photo)

The door handles were of push-button design, while the 1949s had been pull-type. The front piece of the chrome spear bearing the "Custom" designation was redesigned and the gas filler was under a trap door, instead of being exposed.

It's a
two-**FORD** garage
for 250,000 families

Yes, a quarter-million families now own two fine cars—*and both of them are Fords!* These families are enjoying the convenience other people dream about. And they've found that two Fords cost little more than one high-priced car.

Fords are thrifty to buy, thrifty to run, and there's less dollar depreciation at "trade-in" time!

Chances are you are closer to being a two-car family than you think. See your Ford Dealer today and "Test Drive" the two '50 Fords shown here. The car you now own may well provide the down payment on both of them.

The '50 Ford Fordor Sedan. A smart, roomy car that's a pleasure to drive. With King-Size Brakes that act 35% easier. With a "sound-conditioned" Lifeguard Body and Safe-Wide seats. With style that has been awarded the Fashion Academy's Gold Medal 2 years in a row. And with Ford's famous "Mid Ship" Ride to "float" you over the rough spots.

The new Ford "Country Squire" Station Wagon. The "Double Duty" Dandy of them all. Passenger car comfort for eight. New "Stowaway" seat and level tailgate give "Flat Deck" cargo space of 38.5 square feet—more level carrying surface than any other station wagon in its field. (Handles half-ton loads with ease.) A business cargo carrier as well as the family's "fun car."

There's a *Ford* in your future...with a future built in

You travel "FIRST CLASS" without "EXTRA FARE" in the big FORD

Your comfort is "First Class" with Ford's famous "Mid Ship" Ride—cradled between the wheels in the all-steel "Lifeguard" Body! It's a soft ride, a level ride —no jounce, no pitch, no roll! And there is more hip-and-shoulder room on Ford's Sofa-Wide seats than in many cars with upper bracket price tags.

You get "Extra Fare" power from Ford's "whisper-quiet" V-8 and you actually pay hundreds less than others charge for a six! And because of unique "Power Dome" combustion, you get high-compression performance on regular gas!

There's a *Ford* in your future ...with a future built in

Ford gives you "Fashion Car" Styling! From the sweep of its lines and the soundness of its coachwork...to the last smart detail of its "jewel box" interior . . . it's America's best-dressed car!

"Test Drive" the big new Ford! Discover the thrill of driving the one *fine* car in the low-price field! You'll see, hear and feel the difference!

"First-Class" interior comfort was offered in the 1950 Fordor sedan. Ford's famous "Mid-Ship" ride cradled the passengers on "Sofa-Wide" seats between the wheels inside an all-steel "Lifeguard" body.

Ford's first T-bird: 1955

By Robert C. Ackerson

A year after Chevrolet launched the Motorama Corvette, Ford made its response at the Detroit Automobile Show on February 20, 1954. Despite controversy surrounding the fiberglass bodies, the Thunderbird shown at Detroit was a fiberglass mock-up with flat headlight bezels and non-production features.

Ford's eagerness to take the thunder out of the Corvette by giving pre-production Thunderbirds exposure had curious consequences. *Road & Track* displayed a T-bird with 1955 Fairlane headlight peaks on its cover. With a chrome perimeter and simulated air intake, this was an attractive feature, but not one fitted in production.

For a time, it seemed likely that the T-bird would use 1955 Fairlane style body side trim. Ford released a press kit photo of such a T-bird and advertised a car with this trim in *Motor Trend*.

Original specifications called for the 1954 Mercury 256-cubic-inch, 161-horsepower V-8, which was also used as Ford's "Enforcer" police car engine. This would have made an interesting Ford versus Chevrolet contest. With 155 horsepower, the Corvette had a similar power-to-weight ratio. However, the 1955 Corvette's 265-cubic-inch, 195-horsepower V-8 could have been embarrassing to Ford. So, the T-bird came to market with the big 292-cubic-inch V-8, which produced 193 horsepower with standard transmission attachments and 198 horsepower when Fordomatic was ordered.

Having great impact on the T-bird performance image were its achievements at the 1955 Daytona Speed Weeks. In the standing-start mile, T-birds were faster than Jags, Mercedes or Corvettes. The fastest T-bird ran 84.66 miles per hour and was third behind two Ferraris.

The story was the same in the Flying Mile speed runs. A 1955 T-bird's two-way average of 124.6 miles per hour was unequaled by any Jaguar or Corvette. These T-birds ran a three-speed manual transmission with a non-stock 3.33:1 axle combination, but still they made the debate over whether the Thunderbird was a "real" sports car seem silly.

When the T-bird was first displayed, Ford noted that it was the "real thing." The automaker called it "a completely new kind of sports car." Later, Ford adopted the "personal car" label. Uncle Tom McCahill had no trouble regarding the T-bird as a "full-blown sports car," since it was his car that Bill Spear had driven to 124.33 miles per hour at Daytona. *Road & Track* found much to praise and noted, "Ford Motor Company refrains from calling this car a sports car, but we think this policy is overly cautious." *Road & Track* was critical of the changes made for 1956, but called the 1955 model a "touring-sports car ... an extremely practical machine for personal transportation over any kind of distance in any kind of weather." ,

It didn't matter what they called the T-bird, it was popular. Ford made 16,155 in 1955, compared to Chevrolet's 700 Corvettes.

SPECIFICATIONS

Year	1955
Make	Ford
Model	Thunderbird
Body style	two-door convertible
Base price	$2,944
Engine	OHV V-8
Bore x stroke	3.75 x 3.30 inches
CID	292
Compression ratio	8.1:1
Carburetor	Holley 2V
H.P. (Stick)	4400 rpm
Wheelbase	102 inches
Overall length	175.3 inches
Weight	2,980 pounds
Tires	6.70 x 15
OCPG Value	$42,000

The 1955 Thunderbird came to market with Ford's big 292-cubic-inch V-8 producing 193 horsepower with standard transmission and 198 horsepower with Fordomatic. A removable hardtop was standard. (Applegate & Applegate) (Insert: The Thunderbird interior was neat and stylish. A bench seat was included.)

Unlike "real" sports cars, the Thunderbird had roll-up windows. However, it also had performance. At the 1955 Daytona Speed Weeks, in the standing-start mile, T-birds were faster than Jags, Mercedes or Corvettes. (Ford Archives)

When the Thunderbird was first displayed, Ford noted that it was the "real thing." The automaker called it "a completely new kind of sports car." A second version of the hardtop sported porthole windows.

For a time, it seemed likely that the T-bird would use 1955 Fairlane style body side trim. This early proto-type version had this feature and it also was seen in early advertising for the two-seat car. (Ford Archives)

1958 Ford Custom 300

By Gerald Perschbacher

Jackie's daughter, Jane, had come of age. Blooming into a fresh adult life, Jane had just graduated high school with honors. Her mind was set for college, two states away, where she would concentrate on her natural talent in art.

Jackie knew that Jane wanted good transportation and she wanted to surprise Jane with her own car. It had to be a Ford, since Jane's father had nothing but Henry's products since the first Model T hit a dirt road. Since his death, both Jackie and Jane found life a struggle at best, but Jackie had saved enough for at least two years of college and a new Ford.

"Jane, come here," Jackie called one sunny afternoon, after she had returned from the local Ford dealer. "I want you to see something."

Jane had always liked dream books and would spend hours cutting up magazines and pasting down colorful, balanced collages in her scrapbook. "I've got a couple of dream books that are out of this world," Jackie tempted.

Jane rushed lady-like into the living room. Her mother handed her two folders on the new 1958 Fords. "One of your dreams is going to come true Jane; I'm going to give you your own car," said Jackie.

" But mother, how can you afford ...." Jane was cut off in mid-sentence by Jackie's assurance, "Never mind, I can. And I'm determined you shall have a car. Pick the best car in the bunch, Jane, and we'll load it with options, colors and accents you want," Jackie said.

Jane looked for the longest time ... at least it seemed so to Jackie. Silence; not even a nod; then she said, "Mother, I want only basic transportation, nothing frilly. I like my art that way, too. Tasteful, distinctive, classy, but tempered with frugality and common sense. I want this to show in everything I say, do, wear or drive," Jane said.

Jackie was shocked. "But, surely you want some nice extras on a beautiful convertible, Jane." The young woman shook her head. "No convertible; it's not practical for the winters I'll be facing at school."

"Not even a Thunderbird?" asked Jackie.

"Especially not that, even though the name has an artistic ring," replied Jane.

Down the list they went, eliminating station wagons, Fairlanes, coupes and Tudors. Then Jane pointed to the 1958 Ford Custom 300 Fordor sedan. "That's it; here's the one I want," she said in a settled voice.

Jackie looked at the prices the salesman had written in pencil by each car. The Custom 300 Fordor started at $2,246. "Not a high price compared against the $2,433 Fairlane Fordor or the $3,408 Thunderbird," thought Jackie. She reasoned to herself that this left more money for options.

"Okay Jane. If you want the Custom 300 Fordor sedan, it's yours," said Jackie. Now let's put a lot of handy options on it, like these." Jackie rattled off a long list of extras: Ford-Aire suspension at $156; white sidewall tires at $50; back-up lights at $10; front and rear power windows at $101; nine-tube signal-seeking radio and antenna for $99; SelectAire air conditioner with tinted glass for $395; and Interceptor 265 V-8 for $196.

Jane looked unmoved. "Too much, mother; too many frills. It's all froth-on-the-beach; here one minute, gone another. The more fancy things you get, the more trouble because of repairs," Jane rationalized.

She had a point, but Jackie hadn't given up. "At least take power brakes, they're only $37; and you must have power steering for only $69," she said.

"Not even these mother. Only the basics; the six-cylinder engine, six-tube radio and the standard items are enough for me."

"But Jane," her mother said with a disappointed tone. "I so much hoped to make this car a dream car. And I hoped it would be something to remind you of my love and your father's love. He's not here, Jane, but I wanted the car to remind you of him."

"I know mother, and this is why I want it plain. Father took his cars like this every time, and he was proud of it. His simplicity in tastes has given me a fresh outlook on art, something my classwork never will do. I want to be like him ... and you ... full of common sense and good taste. And my Ford Custom 300 Fordor sedan without frills will always remind me of you and father."

SPECIFICATIONS

Year	1958
Make	Ford
Model	Custom 300
Body style	Fordor sedan
Base price	$2,246
Engine	OHV; inline; six-cylinder
Bore x stroke	3.62 x 3.60 inches
CID	223
Compression ratio	8.6:1

Carburetor	Holley 1V
H.P.	145 @ 4200 rpm
Wheelbase	120 inches
Overall length	116.03 inches
Weight	3,222 pounds
Tires	8.00 x 14
OCPG Value	$5,800

Prices for the standard six-cylinder Custom 300 Fordor started at $2,246 in 1958. Back in 1958, this Ford family car represented simplicity in taste, although it looks a bit glitzy today. (Ford Archives)

The 1958 Ford front end was restyled along a Thunderbird theme. A host of factory options and accessories were available for the Ford, but not everyone was used to such items back in 1958. (Ford Archives)

1960 Fords, the forgotten collectible

By Phil Skinner

When the beginning of the 1960s came to America, many bright and wonderful things were predicted for the future. Manned space flights, instant foods, color televisions, transistor radios and passenger jet transports were all on the horizon.

For Ford Motor Company, as well as most of the rest of the auto industry, there was a regaining of balance after a few years of uncertain movements in the sales departments. This was due, in part, to the late-1957 recession.

For 1960, with the exception of the Thunderbird, a totally new line-up of Ford cars was offered to the motoring public. The newest car for Ford was its entry in the compact market, the Falcon. Offered initially as a two-door or four-door sedan, it was the leading sales contender among compacts. Shortly after its October 1959 introduction, the Falcon came out with both a two-door and four-door station wagon, as well as a new Ranchero sedan-pickup.

All 1960 Falcons were powered by a newly-developed six sporting 144 cubic inches and an economical 90 horsepower. In Ford's full-size line-up, there were 17 new models presented. Rarely referred to were the plainest of these cars (which also had the lowest production numbers) the Custom 300 two-door and four-door sedans.

Offered mainly as no-frills fleet cars, the Custom 300s were totally devoid of any extra ornamentation including fender ornaments and F-O-R-D on the hood and rear deck lid. Identified as models 58F for the four-door sedan, and 64H for the two-door sedan, they showed a production of just 572 and 302, respeitively. Often their production figures are shown with those of that year's Fairlanes.

Better known to the regular car-buying public was the Fairlane series. It had been the top-of-the-line in 1956, but was now relegated to the base series for 1960. Available in either two-door or four-door sedans, many of these were used as fleet cars used by the United States Government, police departments, taxi companies and other concerns that ordered such vehicles. A business sedan with the rear seat omitted was also offered in the Fairlane series. It was designed primarily for the traveling salesman.

Next up the ladder were the Fairlane 500s. These, too, were offered as a two-door or four-door sedans, and featured bright window frame moldings, a more attractive interior and extra outside ornamentation. Both Fairlane series shared the same roof line. It featured a large panoramic rear window. This roof line was also shared by the 1960 Edsel sedans and four-door hardtop in their abbreviated final model year. Common to all big 1960 Fords was a new windshield designed to do away with the knee-banging, dog leg "A" pillar.

Having been introduced in mid-1959, the Galaxie series continued on to 1960. The Galaxie, which featured the new Thunderbird-like thick rear "C" pillars, was quite well received this year. It could be ordered in three models: A two-door Club Sedan, a four-door Town Sedan and the only hardtop in the Galaxie line, the four-door Town Victoria.

The Sunliner convertible and the aerodynamic Starliner two-door hardtop were considered to be special series cars for 1960. With no other two-door hardtops in Ford's big-car line-up, the 1960 Starliner sold rather well. Over 68,000 copies were produced and marketed.

Five station wagons were offered this year, plus one commerical station wagon called the Courier. Up until 1958, the Courier had been an enclosed body sedan delivery. In late 1958, and all of 1959, it was basically a two-door wagon without any rear seating. Extra cargo places were provided under the floor board.

Five other wagons consisted of two-door and four-door Ranch Wagons, four-door Country Sedans (six- and nine-passenger) and the famous Country Squire. The latter, with simulated wood siding, was offered only as a nine-passenger unit this year.

The design below the belt line for 1960 was a departure from traditional Ford styling. A long, low hood covered the engine compartment from side to side, opening in a more traditional manner. It was hinged at the rear and made servicing the new Fords more convenient.

The fenders were slab-sided, one-piece units without any top curve. The wheel wells were placed totally under the hood. All of the passenger cars had their series name on the leading portion of the front fender, except the Custom 300s and Fairlanes. These models were devoid of any series script. The Courier, Ranch Wagon and Country Sedan had their series names placed on the lower rear quarter panels, while the Country Squire had it on the rear tailgate.

Leading from the front of the fenders and trailing to the rear of the car was the stainless steel trim known as "The Silver Streak of Success." The rear of the car had fins that were flattened down to be parallel with the ground. In the concave rear panel, which held the center-fill gas door for passenger cars, the Galaxie and special series Starliner and Sunliner had bright trim and back-up lights installed.

On other models, the concave rear panel featured optional back-up lights mounted on body-color sheet metal. What was considered the biggest non-Ford design element were the taillights. These were of half-round shape and mounted on the outer edges of the rear panel. For more nighttime safety, an optional reflector unit (resembling another taillight) could be bolted into the rear bumper in a recessed area.

The taillight theme was carried over into the interior and, particularly, in the instrument cluster. Housed to the extreme left was the fuel level gauge, then the speedometer and odometer, with the temperature gauge to its right. An electric clock was available to occupy the area on the extreme right of the cluster housing. On the Galaxie, Starliner and Sunliner, a chromed series script-bar was placed on the glove box door.

Under the hood were several new ideas. The standard engine was the 223-cubic-inch "Mileage Maker Six" with 145 horsepower. Because of the low hood line this year, it had a new vertically-mounted air filter. One of the more popular optional engines was the 292-cubic-inch Y-block V-8 with 200 horsepower. Available in all models was a 352-cubic-inch, 245-horsepower mill equipped with a two-venturi intake system. For a little more power, the 300-horsepower version of the 352-cubic-inch motor was offered with a four-venturi intake.

Introduced in December 1959 was one of Ford's first entries to the "Total Performance" program. It was a specially-prepared 352-cubic-inch, 360-horsepower engine featuring an aluminum intake manifold, mechanical-advance distributor, specially-cast exhaust headers and unique fuel filter and coil brackets. With this new high-performance package, there were several options that could not be ordered and a couple that had to be. Not available were the automatic transmissions, power steering or power brakes and air conditioning. What had to be ordered was the special handling package, 15-inch wheels (14-inch wheels were standard), heavy-duty brakes and suspension, and electric windshield wipers. Other horsepower-stealing options, such as power seats and windows, were available, but not recommended.

Most of these high-performance packages were installed in the new Starliner. Many of these sleek fastback-styled cars found their way to the winner's circle in those races in which they were entered.

For transmissions, a total of five choices were available for the 1960 Ford. Standard in all models was the manual three-speed. This was followed by the optional three-speed overdrive unit. The two-speed Fordomatic was available behind the 223-cubic-inch six, as well as the 292 Y-block V-8. Cruise-O-Matic was available with the 352 motor in either its 245- or 300-horsepower versions. This was Ford's three-speed, dual-range automatic transmission,

Also new this year was an 11-digit vehicle identification number that represented the year, plant of production, body/model type and engine.

One way in which Ford dealers could jack up the price tag of the 1960 Fords was through the sale of options. Aside from common accessories, such as a radio, heater or air conditioning, many "bolt-on" items were available. Included were an optional hood ornament, bumper guards, the previously-mentioned rear bumper reflectors, swept-back radio antennas and spotlights. Rear bumper "Sports" wheel carriers, commonly referred to as continental kits, were also offered by both Ford and several aftermarket suppliers.

For most collectors today, the 1960 Ford is kind of a forgotten child. Its unusual looks and non-Ford appearance do not appeal to many car collectors. The new Falcon did very well in its debut year and the 1960 Thunderbird had that marque's biggest year up to that time. As if to prove that 1960 wasn't all that Ford Motor Company executives were hoping for, one of Ford's main 1961 advertising themes stressed the point that "traditional styling had returned."

SPECIFICATIONS

Year 1960
Make .... Ford
Model .... Galaxie Special Sunliner
Body style .... Convertible
Base price $2,973
Base engine .... OHV V-8
Bore x stroke .... 3.75 x 3.30 inches
CID 292
Compression ratio 8.8:1
Carburetor .... Holley 2V
H.P. .... 185 @ 4200 rpm
Wheelbase .... 119 inches
Overall length .... 213.7 inches
Weight .... 3,841 pounds
Tires .... 8.00 x 14
OCPG Value $18,000

The Sunliner convertible and the aerodynamic Starliner two-door hardtop were considered to be special series cars for 1960. This is the Sunliner. (Old Cars)

The Fairlane 500s were offered as two-door or four-door sedans and featured bright window frame moldings, a more attractive interior and extra outside ornamentation. (Ford)

With no other two-door hardtops in Ford's big-car line-up, the 1960 Starliner sold rather well. Over 68,000 copies were produced and marketed. (Old Cars)

Ford's 1961 Country Squire

By Robert C. Ackerson

Not until 1961 did Ford offer the Country Squire in six- and nine-passenger forms. This gave Ford six station wagons: two- and four-door six-passenger Ranch Wagons; a pair of four-door Country Sedans in six- or nine-passenger formats and the two Country Squires.

Ford followed its new 1960 models with cars that had all-new sheet metal below the belt line. Although wheelbases remained at 119 inches, overall length was reduced to 209.9 inches. Overall width was also narrowed to 79.9 inches.

There were other dimensional changes as well. The load floor length shrunk from 104.1 inches in 1960 to 99.9 inches in 1961. Load floor width went from 63.8 inches to 62.8 inches. Naturally, the cargo area volume was decreased too. It went from 97.4 cubic feet to 93.5. The rear opening width shrunk over 10 inches, from 60.7 inches to 50.2. Also down in size was the height of the rear opening, which moved from 28.9 inches to 26.3.

Even with the big cut in the width of the Ford tailgate, it was still the widest in its field. It's unlikely that many buyers got their measuring tapes out to measure the difference, but it's very likely that most were impressed with Ford's first use of a power-assisted roll-down station wagon rear window.

With simulated wood trim, the Country Squire was not available in the two-tone paint combinations used on Ranch Wagons and Country Sedans. Still, 13 single colors from Raven Black to Garden Turquoise were offered.

Price-wise, Country Squire V-8s were outranked only by two Thunderbirds in Ford's 1961 line-up. The T-bird coupe and convertible were $4,172 and $4,639 respectively. Country Squires had the following prices: (six-passenger) $2,943 for the six-cylinder and $3,059 for the V-8; (nine-passenger) $3,013 for the six-cylinder and $3,129 for the V-8.

The Country Squire interior had pleated, shimmer-pattern nylon cloth inserts and bright Mylar and Morocco-grained bolsters. Options included air conditioning, power steering, power brakes, heater and defroster, luggage rack, radio, Equa-Lock rear axle, seat belts, power steering, heavy-duty suspension, tinted glass, power windows and two-speed electric windshield wipers.

Base engines for Country Squires were the 223-cubic-inch, 135-horsepower Mileage Maker six and the 292-cubic-inch, 175-horsepower Thunderbird V-8. Both the 352-cubic-inch and 390-cubic-inch Thunderbird Special V-8s were optional with 220 and 300 horsepower. On special order, a Thunderbird 390 Super V-8 with 375 horsepower was available.

Transmission choices were Cruise-O-Matic, Fordomatic and three-speed manual (with or without overdrive). The only combinations unavailable were Cruise-O-Matic with the six and Fordomatic with a 390 Special V-8.

Ford's 1961 wagons contributed 136,619 cars to Ford's total model year output of 1,362,186 units.

SPECIFICATIONS

Year	1961
Make	Ford
Model	Country Squire
Body style	four-door six-passenger station wagon
Base price	$2,941
Engine (base)	OHV; inline; six-cylinder
Bore x stroke	3.62 x 3.60 inches
CID	223
Compression ratio	8.4:1
Carburetor	Holley 1V
H.P.	101 @ 4400 rpm
Wheelbase	120 inches
Overall length	119.9 inches
Weight	3,930 pounds
Tires	8.00 x 14
OCPG Value	$7,000

This ad promoted the "Classic Ford Look For '61." This was the first year that Ford offered six- and nine-passenger Country Squires, which gave it six station wagons.

Power-operated rear window now standard on all '61 Ford 9-passenger wagons and the 6-passenger Country Squire! So handy for loading small items!

All seats face forward. In 9-passenger Ford wagons, everybody rides facing front the way people like to ride! Seat backs flatten flush with the floor for cargo carrying.

Built big for big cargo! And you get a wider rear opening...low loading level...flush tailgate...all make loading and unloading easier for '61.

Beautifully proportioned
to the CLASSIC FORD LOOK

New, for 1961—Ford's famous Country Squire is available in both 6- and 9-passenger models.

BEAUTIFULLY BUILT
TO TAKE CARE OF ITSELF

Beautifully built to do more for you, too, with new seven-inch-wider tailgate, new roll-down rear window and a whole wagonful of built-for-people comfort features no other wagon can match.

A station wagon buyer's guide for '61 would have to put the new Ford Wagon at the top of its list. No other wagon delivers so much, in such style, and at such a favorable price.

In the first place, you choose from America's most complete line ...6 new models...from Country Squire, the Thunderbird of station wagons, to the low priced Ranch Wagon. And there's other important news. Just turning the wheel is 25% easier when parking. New rear springing and wide-tread design give you a smoother ride.

What else is new in the world's most popular wagons? Well, here's a wagon that lubricates itself...cleans its own oil...adjusts its own brakes...guards its own body...takes care of its own finish. In short, a wagon beautifully built to take care of itself!

HERE'S HOW THE '61 FORD TAKES CARE OF ITSELF
...THE FIRST 1961 CAR THAT DEALERS HAVE WARRANTED FOR 12,000 MILES OR ONE FULL YEAR

Lubricates itself—You'll normally go 30,000 miles without a chassis lubrication.

Cleans its own oil—You'll go 4,000 miles between oil changes with Ford's Full-Flow oil filter.

Adjusts its own brakes—New truck-size brakes adjust themselves automatically.

Guards its own muffler—Ford mufflers are double-wrapped and aluminized—normally will last three times as long as ordinary mufflers.

Protects its own body—All vital underbody parts are specially processed to resist rust and corrosion, even to galvanizing the body panels beneath the doors.

Takes care of its own finish—New Diamond Lustre Finish never needs wax.

Extended Warranty—Ford Dealers were the first to extend their warranty on the '61 cars for 12,000 miles or one full year, whichever comes first. Ask your dealer for details.

'61 FORD STATION WAGONS

"Beautifully built to take care of itself," was the theme in this ad. With simulated wood paneling, exterior car care was easy, but Country Squires were not offered with the two-tone finish used on Ranch Wagons and Country Sedans.

1961 Thunderbird unique in all the world

By Peter Winnewisser

We came across a pristine white 1961 Thunderbird hardtop at the Macungie, Pennsylvania car show. Bathed in bright sunlight, it was a real crowd pleaser. The sporty Ford invariably elicited admiring looks and comments from those who walked by.

To our mind, the quicksilver elegance of the 1961 Thunderbird was a forerunner of the aerodynamic lines of 1980s cars. Compare the 1961 model to *Motor Trend's* 1989 "Car of the Year," the Thunderbird SC. Don't you agree that they have much in common?

In 1961, the Thunderbird entered the third stage of its evolution. The first was the 1955 Thunderbird, a two-passenger personal car. The second was in 1958, when it became a four-passenger car.

The 1961 Thunderbird featured elegant new styling, yet it retained the classic concept of the Thunderbird design: the distinctive roof line, four-passenger luxury, the famous center console and the rich interior.

Although it was one-half inch shorter than the 1960 model, its greenhouse was 10 inches longer. This resulted in a more spacious interior. Windshield pillars slanting away from the passengers, combined with wider doors (a full 51 inches wide) to provide a broad opening for easy entrance and exit.

Standard equipment on all 1961 Thunderbirds included the 300-horsepower 390-cubic-inch Thunderbird V-8 engine, Cruise-O-Matic transmission, integral type power steering, self-adjusting power brakes, "Lifeguard" cushioned instrument panel and sun visors, double-grip door locks, safety swivel day/night inside rear view mirror and deep-center steering wheel with horn ring. Also standard were manually-adjustable front individual bucket seats, an electric clock, an automatic dome lamp (in hardtops), a courtesy/map light, dual horns, turn signals, back-up lights and full wheelcovers.

Unique among the many Thunderbird options was a "Swing-Away" steering wheel which, with fingertip effort, slid 10 inches to the right as the driver got in and out. It could be moved to the side or returned when the Cruise-O-Matic selector lever was in "P" (park) position.

The 1961 Thunderbird was available in either the two-door hardtop or a higher-priced ($4,637) convertible. Production totaled 62,535 for the hardtop and $10,516 for the convertible, according to *The Standard Catalog of American Cars 1946-1975*. Ford advertised the 1961 Thunderbirds as "unique in all the world."

SPECIFICATIONS

Year	1961
Make	Ford
Model	Thunderbird
Body style	two-door hardtop
Base price	$4,170
Engine	OHV V-8
Bore x stroke	4.05 x 3.78 inches
CID	390
Compression ratio	9.6:1 (Manual)
Carburetor	Holley 4V
H.P.	300 @ 4600 rpm
Wheelbase	113 inches
Overall length	205 inches
Weight	3,958 pounds
Tires	8.00 x 14
OCPG Value	$13,000

The quicksilver elegance of the 1961 Thunderbird was a forerunner of the aerodynamic lines of 1980s cars. The hardtop had a handsome profile and was the car to be seen in at your local riding stable.

In 1961, the Thunderbird entered the third stage of its evolution. First came the 1955 two-passenger, then the 1958 four-passenger. The new design was obviously geared to the "swinging '60s" lifestyle.

The 1961 Thunderbird hardtop featured new styling, but kept the T-bird's distinctive roof line, four-passenger luxury, center console and rich interior. The large round taillamps were a Ford trademark.

Perfect for a romantic evening at the beach was a red Thunderbird convertible. Standard equipment on all 1961 Thunderbirds included the 300-horsepower 390-cubic-inch Thunderbird V-8 engine.

Hudson's Country Club for 1939

By Robert C. Ackerson

Back in 1939, Hudson salesman had plenty to tell customers, whether they were long-time loyalists or first time prospects. Hudson could boast of new styling for all its cars, except for the 112. It had been introduced in 1938 and received only a minor face-lifting its second season.

Although the line-up did not include the Terraplane (called Hudson-Terraplane the last year), it did feature a new Country Club line. Hudson described them as being positioned in "the moderate price field." These cars, along with the less expensive Hudson Six, displayed most of Hudson's best known virtues and set the theme for Hudson styling until the arrival of the legendary Step-Down in 1948.

While it reflected Lincoln-Zephyr's styling leadership, rather than a trail-blazing design move, the new Country Club front end was accurately depicted as having a "modern, low, long, fleet appearance." The headlights were mounted flush into the fenders and the die-cast center grille had very slender horizontal bars. Two smaller die-cast grilles were positioned in the catwalk region of the front fenders. A touch of class and distinction was provided by the Country Club's parking lights, which were placed at the front end of the hood moldings.

Hudson claimed the interiors of its 1939 models provided the "greatest degree of luxury ever incorporated into its line." Standard equipment on Hudson Sixes and Country Clubs included an electric clock, ashtrays, armrests, sliding rear quarter windows and crank-controlled front ventilator wings.

Country Clubs were available, with six- or eight-cylinder engines, on a 122-inch wheelbase. The Custom Touring Sedan had a 129-inch wheelbase. A feature unique to it was front window deflectors that could roll down into the doors. The 122-inch wheelbase Country Clubs with closed bodies included a two-passenger Victoria Coupe, a two-passenger Touring Brougham, a four-passenger Touring Sedan and a two-passenger Coupe.

The upholstery in these Country Clubs was cashmere cloth. A fawn gray Bedford cord with leather and chrome trim was standard on the Country Club Custom Touring Sedan. The Convertible Brougham and Coupe had hand-buffed leather upholstery.

Auto manufacturers of the late 1930s were eager to have the public associate their products with trains or airplanes. Hudson's new "Air-foam" seat cushions made such a connection. "Cushions made of the same material have been successfully used on the latest streamline rail trains and on crack transcontinental airliners," said sales literature. Essentially, the seats were formed of latex rubber and filled with inter-connected bubbles that permitted air to flow through the cushion as it moved up and down.

Another feature that Hudson claimed provided improved cooling and ventilation was the perforation of the seat cushion with thousands of needles during its manufacture.

Hudson continued to use semi-elliptic springs at both the front and rear in 1939. This setup was refined by use of a transverse-mounted front torsion bar, which was attached to the wheel spindles. Hudson called it "Auto-Poise Control." It was a simple, but effective method of improving ride stability. Whenever the wheel was turned away from center, the twisting of the bar tended to pull it back.

Hudson was justifiably proud of being rated "America's Safest Automobile" by the National Safety Council in 1938. The use of Auto-Poise Control, a forward-hinged hood, "Double Safe" brakes with both mechanical and hydraulic functions, as well as the availability of back-up lights and a rear window wiper system were assurances that this reputation would be further enhanced.

All Hudsons had column-mounted shift levers and the Country Club models could be ordered with Selective Automatic Shift, which provided electrically-controlled vacuum shifting. This setup could also be linked to an automatic clutch.

1939 wasn't a record production year for Hudson, yet the output of 82,161 cars was a respectable achievement. One of these vehicles was a Country Club Touring Sedan, built on May 4 with serial number 2,614,165, which was selected to honor Hudson Motor Car Company's 30th anniversary.

SPECIFICATIONS

Year	1939
Make	Hudson
Model	Country Club 8
Body style	Touring Sedan
Base price	$1,079
Engine	Inline; L-head; 8
Bore x stroke	3.00 x 4.50 inches
CID	254
Compression ratio	6.25:1

```
Carburetor...........................................................................................Carter WDO 2V
H.P. ...............................................................................................122 @ 4200 rpm
Wheelbase .......................................................................................... 122 inches
Overall length ..................................................................................... 199 inches
Weight ............................................................................................ 3,193 pounds
Tires .................................................................................................. 6.25 x 16
OCPG Value ....................................................................................... $13,500
```

The 1939 line-up had no Terraplane, but added the new Country Club Touring Sedan. While it reflected Lincoln-Zephyr's styling leadership, the new Country Club front end was accurately depicted as having a "modern, low, long, fleet appearance."

1954 Imperial:
Eisenhower era exclusive

By Dave Duricy, Jr.

Pan American Road Race Lincolns, Cadillac Eldorados and Packard Caribbeans have all had their share of exposure. No discussion of early- to mid-1950s luxury automobiles could take place without their mention, but no such discussion would be complete without a nod to the 1954 Imperials.

Luxurious, imposing, lavishly powerful and lacking in a certain refinement, the 1954 Imperials are very much creatures of the Eisenhower era. Their appeal lies within the aura of strength and confidence that seems to surround them.

Clad in an elegant evolution of "three-box" armor, the 1954 Imperials were the ultimate expression of the styling concepts that Chrysler had been working on since the introduction of its all-new 1949 models. The look consisted of clearly defined rear fenders that emerged from the body sides and jutted cleanly rearward, creating the first box. The second box was created by a strong beltline that led from the headlight to the leading edge of the rear fender. The greenhouse, hood and trunk surfaces composed the third box.

Despite what the term three-box might lead one to visualize, the 1954 Imperials were not angular and upright creations with sharp creases and flat, characterless expanses of sheet metal. The enormous hoods, topped by impressive brightwork eagles with wings thrust backwards, sloped gently down and forward to meet the grilles. All fenders had rounded edges and a slight outward curve that gave the body a look of fullness and masculinity. The effect is formal, though with a certain dash of sportiness.

The Chrysler-based Imperials looked more affluent than their Lincoln counterparts and certainly saw eye-to-eye with Packard in the area of how a car should look. Cadillacs from the era, when compared to the restrained Imperials, look faddish and trendy with their odd appendages, massive chrome bumper bombs and dog-legged wraparound windshields.

The 1954 model line appealed to an exclusive clientele, one that shunned ostentation, but appreciated flair and respected fine workmanship and design. Many also had a need for chauffeur-driven transportation.

The series contained no fewer than three limousines. Two made up the Crown Imperial sub-series and rode on regal 145.5-inch wheelbases. The third was found in the Custom sub-series. It was a town car designed to chauffeur the lady into town so she might run her errands.

Owner-driven Imperials were limited to just two models, which were a truly good-looking Imperial Custom Newport two-door hardtop and a four-door Imperial Custom Sedan.

The sedan rode on a 133.5-inch wheelbase which was eight inches longer than that used for the Chrysler New Yorker. The Newport hardtop's wheelbase measured six inches longer than its lesser Chrysler counterparts, which had 125.5-inch spans. These standard Imperials were leviathans with stem-to-stern measurements of about 19 feet.

To propel this mass, all Imperials were equipped with the 331.3-cubic-inch hemi V-8 rated at 235 horsepower. This power plant, shared with the New Yorker Deluxe, made the 1954 Imperials the most powerful automobiles in their class. The most oomph Packard could muster was 212 horsepower from an old-fashioned (though smooth) straight eight. Lincoln delivered 205 horsepower from its V-8, while Cadillac nipped at Imperial's heels with a 230 horsepower V-8.

The modern, efficient hemi V-8 made the 1954 Imperials appealing indeed, especially coupled with the fact that all models were equipped with PowerFlite automatic transmission. Those in the market for limousines must have been impressed with the Crown Imperials' standard power plant and standard disc brake system. Not only did the Imperial come with the most advanced braking system in the industry, but Chrysler provided power steering as well.

Inside, Imperial drivers and passengers found generous quantities of high-quality upholstery and carpeting. Bright chrome accents were used liberally, creating a bejeweled environment that featured gigantic, sofa-style, "chair high" seats.

The entire 1954 line sold in exclusive numbers. A mere 77 Crown limousines were produced. Its six-passenger companion saw only 23 copies made. Volume for the Custom Sedan was a healthier 4,374 units and the lovely Custom Newport found 1,249 buyers. The Custom limousine was the most popular limo produced by Imperial, with only 85 built. These are truly limited numbers. This means that 1954 Imperials were (and still are) rare sights.

Unlike Cadillac, whose total production exceeded that of American Motors and Studebaker, the 1954 Imperial wasn't a car that could be owned by just anyone. By comparison, even Lincoln was a common make with its 35,733 assemblies.

1954 was the last year for the Imperial as a member of the Chrysler line. For 1955, it would be marketed as a separate nameplate. This, plus many attributes, make 1954 Imperials special reminders of early 1950s luxury motoring.

SPECIFICATIONS

Year	1954
Make	Chrysler
Model	Imperial
Body style	Custom Newport two-door hardtop
Base price	$4,560
Engine	OHV V-8 (hemi)
Bore x stroke	3.81 x 3.63 inches
CID	331.1
Compression ratio	7.5:1
Carburetor	Carter WCFB-2041S 4V
H.P.	235 @ 4400 rpm
Wheelbase	131.5 inches
Overall length	221.75 inches
Weight	4,345 pounds
Tires	8.20 x 15
OCPG Value	$16,000

The "three-box" styling was the ultimate expression of the concept that Chrysler had been working on since the introducing its all-new postwar models. Imperials were based on New Yorkers with Derham Body Company upgrades.

1961 Imperial: The car that time forgot

By Dave Duricy, Jr.

Model year 1961 was important for American automotive design. It was the year that Lincoln discovered its identity and at the same time established the automotive fashion of the 1960s. It was the year of the Continental, the four-door that dared to be plain.

No fins, no wildly sculptured body panels, no globs of chrome or funky headlights, the new Lincoln demonstrated that restraint could be fashionable. Sales jumped by over 10,000 units and the look was all the rage. The rest of the luxury car industry was caught off guard.

Perhaps, the most surprised was the Imperial, for the 1961 model line was the most brazen the nameplate had ever seen. Head stylist Virgil Exner had let his imagination run wild when he planned the 1961. Toying with Classic-era styling motifs such as free-standing headlights and tall clamshell fenders and sweeping lines, Exner created a car that was everything that the Lincoln wasn't. It was big, bold and soon to become dated.

Even so, the Imperial was an exciting package. The three owner/driver model lines ... Custom, Crown and LeBaron ... were all powered by the most powerful V-8 in the luxury car field. It was the 413-cubic-inch wedge head engine packing 350 horsepower. The power generated by this giant engine was transferred to the wheels through a TorqueFlite automatic transmission. The driver had command of the power flow via a bank of push-button transmission controls. A click of the drive button and a stomp of the accelerator hurled the finned giant down the road.

Chrysler argued, during the 1950s, that tailfins helped a car's high speed stability. If so, the 1961 Imperial must be the most stable freeway car ever produced, as its fins were the largest ever worn by an Imperial. Regardless of just how functional the fins were or weren't, they certainly added to the 1961 design. Angled up and brought to a sharp point, the great fins gave the car a look of distinction and speed.

The startling rear end was matched by the equally dramatic front. It was here that Exner's infatuation with Classic-era automobiles made itself most visible. Pockets beneath the leading edge of the front fenders, meant to resemble the free-standing fenders of the 1930s, harbored sets of free-standing headlights. Nestled in chrome pods and set upon tiny chrome supports, the headlights were the most unique element of an already unique design. These wild headlights were positioned alongside a classically-inspired upright grille, which was complete with a driver's side badge bearing a rendition of the handsome Imperial eagle.

The silhouette of the car was typical Exner. The angle of the roof line, the cant of the grille and the height of the fin all combined to give the car a lunging appearance. Accentuating this quality was a full length chrome spear that swept from the brow of the headlights to the rear of the fin. Even the curve of the wheel cutouts added to the "Forward Look."

Perhaps the most elegant looking of the 1961 Imperials was the LeBaron. Unlike other Imperials, which had narrow rear roof pillars and large window areas, the LeBaron was equipped with a formal, thick rear pillar. Not only did the special roof add rear seat privacy, but a certain limousine-like charm as well.

Of course, Imperial was still producing limousines in 1961 and the model year's was one of the marque's most impressive. Factory price was $16,500 for the 149.5-inch wheelbase, 6,000-pound Crown Imperial. It was equipped to the hilt with luxury features such as air conditioning, power windows, automatic headlight dimmer, cruise control and three heaters. Production was just nine cars.

The mid-line Imperial Crown saw a much healthier production total. In fact, it was the most popular model of the 1961 Imperial line. It was offered in two hardtop body styles, two-door and four-door, plus a convertible and found over 6,000 buyers. Nicely equipped, the Crowns offered a vanity mirror, power windows and a six-way power seat, just to name a few standard features. Of course, power brakes and power steering were standard for all Imperials.

Despite the many pleasing aspects of the 1961 Imperial, it has been the object of much criticism over the years; its styling condemned as a vulgar example of Exner's "last stand."

To demonstrate its excess, the Imperial is often compared to its Lincoln counterpart, with the Lincoln hailed as a design triumph of almost mythical proportions. However, in this age of look-alike cars and black trim, the Imperial is a delight.

Examining Imperials' 1961 design reveals a cohesive and "classic" character. The 1961 Lincoln may have been representative of the styling wave of the future, but the 1961 Imperial represents a past styling age now sorely missed.

SPECIFICATIONS

Year 1961
Make...Imperial
Model...Crown
Body style..four-door hardtop
Base price...$5,649
Engine ....OHV V-8
Bore x stroke..4.18 x 3.75 inches
CID.. 413.2
Compression ratio...10.1:1
Carburetor..Carter AFB 4V
H.P. ....350 @ 4600 rpm
Wheelbase...129 inches
Overall length ....227.1 inches
Weight ....4,855 pounds
Tires ....8.20 x 15
OCPG Value $7,500

The 1961 Imperial was the most brazenly designed car the nameplate had ever graced. Head stylist Virgil Exner had let his imagination run wild. The free-standing headlamps were thought to have a "classic" look. (Chrysler photo)

In this age of look-alike cars that wear flat-black trim, the 1961 Imperial is a delight. It's shimmering and exciting. The Custom, Crown and LeBaron all had the 413-cubic-inch wedge V-8, most powerful engine in the luxury field.

The 1961 model's shark fins were the largest ever worn by an Imperial in the history of the marque. The "gunsight" taillamps had a stuck-on appearance. Another unusual styling treatment was the wrapover roof.

1962 Imperial challenged the luxury car market

By John A. Gunnell

In its 1962 advertising, the Imperial was pitched as "America's most carefully built car." Chrysler announced, in one ad, an invitation for potential buyers to arrange a "comparison tour" during which they could ride in the Imperial and see how it stacked up against its competitors.

The automaker claimed this was not the usually vague "see your dealer" hyperbole, but rather a "forthright challenge" to try the best Imperial ever built. "We believe anyone who is planning to invest from $5,000 to $7,000 in an automobile is entitled to the unembroided facts right from the car itself," said the copywriters. "So, when our invitation arrives, take us up on it."

As you may recall, 1962 was the year when Chrysler's towering tailfins started to come back down towards earth. The Imperial followed the trend, converting the previous shark fins into an integrated, forward-tapering fender contour. There was a new hood ornament and a revised grille with thin horizontal bars in a split-section design. Free-standing taillamps, mounted atop the new wedge-shaped fenders, complimented the classic car-inspired headlamps-on-pedestals of 1961.

The car depicted in the advertisement was the Imperial LeBaron four-door Southampton model. The top-of-the-line LeBarons had their own distinctive rectangular rear window and formal-styled roof line. In Chrysler lingo, "Southampton" stood for hardtop, so this was a pillarless sedan.

While larger and heavier than all other Imperial models (there were no limousines made in 1962), the LeBaron shared the same big V-8 with other series and also shared the advantages of a torsion-bar suspension, an electrical system alternator and a newly improved TorqueFlite push-button automatic transmission.

The 1962 Imperial LeBaron Southampton took its first bow on September 26, 1961, but made only a minor contribution to the Imperial car-line's 14,337 production units that model year. In fact, only 1,440 were made. That wasn't too bad, however, as there were three other Imperials that had smaller runs. The Imperial Custom two-door Southampton hardtop found 826 buyers, while the same body type in the Crown Imperial series saw 1,010 assemblies. Rarest car of the year, though, was the Crown Imperial convertible, of which only 554 copies were created.

Marketing-wise, the big news for 1962 was the transfer of the Imperial car-line to the Chrysler-Plymouth Division. In 1955, the Imperial name had been made a separate marque (then considered a product of the Imperial-New Yorker Division of Chrysler Corporation). Now it was returning to a somewhat similar status. Of course, this had nothing to with the product itself. It was simply too expensive to maintain the operation of a separate marketing arm for a series with such a low sales penetration level.

Today, collectors appreciate all of the original features of the 1962 Imperial. In addition, they have even greater appreciation for its scarcity. That's what helps to make it a rare and valuable marque.

SPECIFICATIONS

Year	1962
Make	Imperial
Model	Southampton
Body style	four-door hardtop
Base price	$6,422
Engine	OHV V-8
Bore x stroke	4.18 x 3.75 inches
CID	413.2
Compression ratio	10.1:1
Carburetor	Carter AFB 4V
H.P.	340 @ 4600 rpm
Wheelbase	129 inches
Weight	4,725 pounds
Tires	8.20 x 15
OCPG Value	$10,000

IMPERIAL LEBARON FOUR-DOOR SOUTHAMPTON

A forthright challenge to everyone who plans to buy a luxury car this year

In a few days, you will receive, by mail or telephone, a personal invitation to drive a 1962 Imperial.

Not the usual vague "see your dealer" . . . but a specific challenge to compare your own car with the best Imperial we've ever built.

At your convenience, a dealer in your area will deliver a new 1962 Imperial to you . . . for a thorough comparison-tour that you conduct all by yourself.

We believe anyone who is planning to invest from five thousand to seven thousand dollars in an automobile is entitled to the unembroidered facts right from the car itself.

Naturally, our dealers will gladly *explain* the advantages of torsion-bar suspension . . . *tell* you how the alternator supplies electricity even while the engine idles . . . *quote* engineering data on the superior performance of our new transmission . . . and answer any other question you ask about Imperial.

But you can find out only so much about Imperial by mere *listening*. To give the facts substance and meaning you must *drive* and *compare*.

So, when our invitation arrives, take us up on it. It obligates you not at all. And whether you eventually buy an Imperial or not, you'll never forget that once you drove a car which handled and accelerated and thrilled as a great car is supposed to.

IMPERIAL
AMERICA'S MOST CAREFULLY BUILT CAR

R.S.V.P. *Even though our invitation may somehow miss you, an Imperial comparison-tour may easily be arranged by writing on your letterhead to: General Manager, Imperial Division, 12200 East Jefferson, Detroit, Michigan.*
IMPERIAL — A PRODUCT OF CHRYSLER CORPORATION

In this 1962 advertisement aimed at upscale buyers, the latest Imperial LeBaron four-door Southampton (four-door hardtop) was pitched as "America's most carefully built car."

IMPERIAL CROWN FOUR-DOOR SOUTHAMPTON

TO AMERICA'S 5,344 LEADING M.D.s

In a few days, you (and a number of your medical colleagues) will receive a letter or a phone call offering you the personal use of a new Imperial.

The car will be delivered to you. Our representative will point out the various controls, and then it's yours . . . for a thorough comparison-test.

Our dealers are well equipped to discuss the advantages of our push-button transmission, our torsion-bar suspension. They know Imperial has the largest brakes and most powerful engine of any American fine-car. But engineering facts can only suggest its crisp precision, its faultless smoothness, its astounding road performance.

Therefore, we offer you a privately conducted test . . . to compare Imperial objectively with other cars you've driven . . . a preview of what great cars are supposed to be, and do.

Please accept our invitation. There's no obligation . . . we simply want to be sure you don't overlook anything in selecting your next fine-car.

Even though you aren't a doctor, we'll gladly arrange an Imperial comparison tour for you. Write on your letterhead to: General Manager, Imperial Division, 12200 East Jefferson, Detroit, Michigan

IMPERIAL
America's Most Carefully Built Car

IMPERIAL — A PRODUCT OF CHRYSLER CORPORATION

Here's another 1962 advertisement that was aimed at the country's 5,344 "leading doctors." This was the year that Chrysler's towering tailfins came back down towards earth and the Imperial followed the trend.

IMPERIAL CROWN FOUR-DOOR SOUTHAMPTON

LEADING DOCTORS ACCEPT NEW IMPERIALS FOR COMPARISON TESTING

Recently we invited the nation's leading doctors to drive new Imperials. Our dealers are continuing to deliver these cars to the doctors' personal care for a thorough comparison tour.

We have urged them to make a thorough diagnosis of Imperial's manner of handling, its performance, its over-all elegance and quality.

Our guests express surprise that a car of Imperial's size can handle so effortlessly. Those who own other fine cars are particularly impressed with Imperial's greater degree of comfort and over-all performance.

These are precisely the major points these tests were designed to prove.

It may be you are *not* a doctor, but would like to discover for yourself the true meaning of Imperial's hearty performance, its splendid torsion-bar balance, its effortless power-assisted steering and braking.

If you would like to enjoy a personal test of America's finest fine car . . . write on your letterhead to: General Manager, Imperial Division, 12200 East Jefferson, Detroit, Michigan.

IMPERIAL
America's Most Carefully Built Car

IMPERIAL—A PRODUCT OF CHRYSLER CORPORATION

This ad advises that the "leading doctors" (it doesn't say how many) had accepted Chrysler's offer to test the Imperial. There was a new hood ornament and a revised grille with thin horizontal bars in a split-section design.

After the doctors came the lawyers. Chrysler also "proposed" an Imperial test drive to them. The 1962 models bowed on September 26, 1961. Only 1,440 Southampton sedans (the model featured in most ads) were made.

1954 King Midget:
little, but long-lasting

By R. Perry Zavitz

Countless new would-be carmakers sprang up after World War II. Most disappeared before we ever saw their cars.

One manufacturer, however, declared in 1953, "Henry Kaiser and we are the only ones out of the crowd who were able to hang on." Two years later Kaiser disappeared to South America, but the other survivor remained for 16 more years. Who was this carmaker? Ever hear of King Midget?

It was the tiny car advertised in those tiny ads in *Popular Mechanics* and similar magazines. Some people might object to calling this forerunner of the go-cart a car. But, for now, let's give this little fellow the benefit of the doubt.

During the war, Claud Dry of Oklahoma and Dale Orcutt of Indiana flew together in the Civil Air Patrol. They convinced themselves that a small, light car could be built at an attractive price to serve a need no one bothered to meet. After the war, they settled in Athens, Ohio, a city south of Zanesville.

In 1945, they established Midget Motors Manufacturing Company. While Dry worked as a linotype mechanic and Orcutt a machinist in a clothing factory, they spent all their spare time setting up the factory in a converted supermarket. Their initial production was motor scooters, but by mid-1947, cars began coming out of their makeshift factory.

The first King Midget was a single-seater that looked like a contemporary quarter-midget racer with headlights and cycle fenders. It had a single-cylinder, rear-mounted, air-cooled, six-horsepower Wisconsin engine and a single-speed automatic transmission.

When the engine idled, the clutch was disengaged. As the throttle (a push/pull button on the dash instead of a foot pedal) was opened, the engine revved and the clutch would engage. The engine drove just the right rear wheel. Reverse was twice is reliable as any other, according to Tom McCahill, who explained how it operated. "You hoist your leg over the side and push backward with your foot!"

Forget 0 to 60 times, because the King Midget had none. Top speed was almost 50 miles per hour, depending on the driver's nerve. Of course, that was a guess, since there was no speedometer. But, King Midget emphasized its fuel economy, sometimes reaching 80 or 90 miles per gallon.

Unfortunately, the factory structure necessitated sub-assembly work to be done on the ground floor, with final assembly on the second floor. Finished, the cars were too big (if you can believe that) to be carried down the narrow stairway. So, they had to be swung out a window and lowered to the ground by a winch. Such awkwardness prompted the company to sell their cars in kit form, to avoid the hassle.

The earliest models were priced at $270 for the knocked-down unit. Consequently, the King Midget was advertised as the "world's lowest priced car." Earlier, of course, Henry Ford sold some of his Model T at about that price. But, Ford wasn't a King Midget kind of car. The Model T was not a postwar car. Nor did it have the luxury of single-passenger exclusiveness. The Model T buyer was denied the chance to assemble his own car.

King Midget's assembly line, instead of moving like Henry Ford's, stood still. Several chassis were laid out on the floor and the workers then attached the parts to them in those set locations.

Midget Motors had no dealer network. Sales were largely generated by those little ads in *Popular Mechanics*. Old issues brought in nearly as many sales orders as current issues. Word of mouth was the greatest source of orders. A "Rider Agent" plan was established. It paid owners a commission on each sale made. Four or five orders paid for the cost of the initial car.

Orders flowed in at a satisfactory rate. Dry and Orcutt were not interested in becoming a big operation. Their staff of 20 assemblers were kept busy producing King Midgets at the rate of around 150 a month.

The company moved into an older, but more suitable, building in 1948. In 1951, a new model was put into production. It had a bigger engine. The 23-cubic-inch air-cooled, Wisconsin one-lunger was rated (conservatively they said) at 7-1/2 horsepower. By 1953, an 8-1/2 horsepower engine was used.

The 25 percent or 42 percent power boost was needed because seating capacity had doubled to two. However, all that extra power came with a price. Fuel consumption plummeted to just 65 miles a gallon.

A new two-speed automatic transmission was featured in this new King Midget. Designed and patented by Dale Orcutt, it had only 75 parts, compared to 1,100 in a normal automatic. And he didn't forget to put a reverse in it this time.

Suspension was another patented innovation. Each wheel had its own coil spring in an oil bath, which was integrated with an oil and air shock absorber, also patented.

The new two-passenger King Midget came with such luxuries as a windshield and a cloth top, which owners of previous King Midgets were not allowed.

Replacing the former wood frame was an aircraft-type perforated girder-and-tube affair. It was light enough for one man to lift. Yet, it was strong enough to bear the weight of 20 King Midget workers, as factory pictures show. The King Midget claimed to be the only car capable of carrying more than its own weight.

Tipping the scales at 460 pounds, the 1951 King Midget cost under $600. Wheelbase was a scant 72 inches and overall length was 96 inches. In other words, the King Midget was so small it could be parked between the axles of standard-size cars.

The next generation King Midget was the 1959 model. Though it had angular fenders like its predecessor, the body was different from its forward-slanting front to its almost fin-like taillight mountings.

The Wisconsin engine was given a boost to 9-1/4 horsepower. The matchless two-speed automatic transmission remained. New hydraulic, four-wheel, self-equalizing brakes were featured. Also doors and side curtains were available.

This was the last model change for King Midget, but improvements were always incorporated when developed. In 1967, a bigger, more powerful engine was installed. King Midget switched from Wisconsin to Kohler. Bursting with 29 cubic inches, this King Midget boasted 12 horsepower, double the power at the beginning.

Just like the big cars, King Midget got bigger engines and more power. The car itself was bigger. Now, it was 117 inches long. Its prices also grew. By 1968, it cost over $1,000.

Dry and Orcutt sold out in 1966, but stayed on as consultants. They sold at a good time, because the King Midget's fortunes began to fade. The last King Midget was built in 1969. Total production is uncertain, but it was in excess of 5,000.

Aside from all its innovations, the King Midget's greatest achievement was its longevity. Because it had no direct competition and because it offered quality and value, it enjoyed an amazingly long existence. It was in production for 22 years.

SPECIFICATIONS

Year	1954
Make	King Midget
Body style	Roadster
Base price	$550
Engine	One-Cylinder; L-head; Air-cooled
Bore x stroke	3.00 x 3.25 inches
CID	23
H.P.	7.5 at 3600 rpm
Wheelbase	72 inches
Overall length	102 inches
Weight	450 pounds
Tires	4.00 x 8
OCPG Value	$3,200

The 1954 King Midget was powered by a 23-cubic-inch, 8-1/2-horsepower engine. Later, 9-1/4-horsepower Wisconsin and 12-horsepower Kohler engines were used.

In addition to the Lincoln-Zephyr ragtops, there was also a 1938 Lincoln Model K convertible coupe with totally different styling. While very attractive, this sidemounted version isn't as modern-looking as the Zephyr.

Lincoln's new "teardrop" ragtop for '38

By Tim Howley

The Lincoln-Zephyr was a radical departure from the famous K series Lincolns. It was introduced for the 1936 model year as Ford's first entry into the medium-priced field. For 1936, there was only the two-door and four-door sedans. A Coupe-Sedan and Town Limousine were added in 1937. "Two racy new convertibles, a sedan and a coupe, bring the total number of bodies up to six," said *Motor* in its show number (November 1937) describing the 1938 line-up.

The convertible coupe had originally been proposed by Lincoln-Zephyr designer John Tjaarda for 1936. Since a convertible sedan was the common companion to the convertible coupe in those days, it made sense to add one of each style to the 1938 Lincoln-Zephyr series. They were probably intended to keep the company competitive with several other manufacturers selling cars in the same price range.

There were actually three distinct series of 1938 Lincoln-Zephyr convertible coupes. Only cars in the third series were five-passenger versions with a back seat. The three series were offered sequentially, not simultaneously, so the third series was the final one. In all, only 600 of these 1938 Lincoln-Zephyr convertible coupes were produced. Judging by cars that turn up at shows, they have a slightly better survival rate than 1938 convertible sedans, of which 461 were manufactured.

All of the Zephyrs underwent their first major facelift in 1938. A totally revised front end treatment identified the 1938 models. Described by *Motor* as "a brand new and attractive idea in styling," it was called a "tear-drop" front end. The customary radiator grille was discarded by a nose consisting of gracefully formed panels and air passed to the radiator through two grille openings in the base of the design. The grilles had multiple, thin, curved horizontal bars. Some collectors seen to prefer the 1936-1937 appearance, while others see the revised design as Lincoln-Zephyr styling at its "art deco" best.

There was an increase in wheelbase from 122- to 125-inches and an accompanying increase in the rear spring base for an improved ride. This also allowed the engine and transmission to be shifted forward for an increase in the length of the hood and the depth of the driver's compartment. The transmission tunnel was lowered ("practically eliminated," said *Motor*) and the gear-shifting mechanism was relocated in a strange manner. The central tower which carried the instruments at its top, plus the heater and radio speaker, concealed the gearshift lever, except for a horizontal extension of the lever towards the rear. This brought the shift knob close to the steering wheel and created a really bizarre (though not necessarily more convenient) arrangement.

At the rear, the height of the drive shaft tunnel was also reduced 1.3 inches through the adoption of a hypoid rear axle and elimination of "straddle mounting" of the pinion. Up front, a lower and wider radiator was used with the new grille design and fan efficiency was upped to help maintain adequate cooling.

The frame was strengthened considerably. Side rails used on closed cars had a 1/2-inch deep section and consisted of two channels, one inside the other, that were welded along the edges, as well as to the steel body. This made the whole structure a single-unit design. On the two new convertibles, the frame side rails also had a plate welded to the open face of the channel to form a complete box section. There was also a stiff X-member consisting of similar dual boxed-in channels. As a further aid to rigidity, the door posts were reinforced with triangular plates that were welded to the frame.

What many people dislike about the 1938 Lincoln-Zephyr is its brakes. Thankfully, this was the last year of mechanical brakes for Ford. That means that the 1938 convertible coupe and convertible sedan are the only open-bodied Zephyrs to have mechanical brakes.

Several engine advances were aimed at quieter operation. They included hydraulic valve lifters, a new combustion chamber shape, and compression-cup type engine mounts. The V-12's horsepower remained at 110 and the displacement stayed at 267.2 cubic inches (through 1939). The belief still persists to this day that the V-12 was an oil hog and just a lot of trouble. This was only true if you lugged the car. Actually, it was a real high-performance engine and, consequently, Lincoln-Zephyrs found a lot of popularity among salesmen who changed the oil regularly and drove the cars 100,000 miles or more.

Engine problems usually came in with second owners who bought these cars with over 100,000 miles on them. Remember, this was before odometer roll-back laws. The buyers had no way of determining true mileage. They were often under the impression that they were buying 20,000- to 30,000-mile cars. Little did they know!

Other mechanical changes for 1938 included the use of angle-mounted rear spring shackles in place of the transverse stabilizing rod and vertical shackles used in 1937. The transmission had a new "blocker" type of synchronizer and there was a new 18-inch diameter steering wheel. The battery went under the hood and a voltage regulator was added to the generator.

Inside, the Zephyrs were also newly styled. They had new seat cushions with small diameter, closely nested coil springs. The instrument panel was finished in a soft-toned beige color with upholstery fabrics, moldings and interior hardware selected to match. A new instrument panel was of a rounded form with recessed controls and gauges grouped in a single cluster at the center. The controls were arranged in a semi-circle around the instruments. There was an ashtray at each side, but the left-hand one was removable for installation of the radio dial and controls.

Exterior hardware was all-new, with redesigned hubcaps and a combination taillight/stoplight recessed into the rear fenders. The license plate holder and lamp were located in the center of the rear deck lid. All bodies were finished in baked enamel.

SPECIFICATIONS

Year	1938
Make	Lincoln
Model	Zephyr
Body style	convertible coupe
Base price	$1,650
Engine	L-head; V-12
Bore x stroke	2-3/4 x 3-3/4 inches
CID	267.3
Compression ratio	6.7:1
Carburetor	Chandler Grooves AA1 2V
H.P.	110 @ 3900 rpm
Wheelbase	125 inches
Overall length	210.9 inches
Weight	3,489 pounds
Tires	16 x 7.00
OCPG Value	$40,000

Two racy new convertibles brought the total number of Lincoln-Zephyr models up to six in 1938. The convertible coupe was sporty-looking, while this convertible sedan exuded class and grace. (Ford Archives)

These bird watchers traveled in style. Since a convertible sedan was the common companion to a convertible coupe in 1938, it made sense to add one of each style to the Lincoln-Zephyr series. (Ford Archives)

1951's "small" Lincoln

By Bill Siuru

Today, almost identical sheet metal is routinely shared by the U.S. automakers for several of their marques. Such cars are now being called "clones." It is next to impossible to tell some Chevrolets from some Oldsmobiles, some Dodges from Plymouths, or some Mercurys from Fords.

Back in the 1940s and 1950s, most cars had rather characteristic and easy-to-recognize features. However, there were some exceptions. Among them were the "small" 1949-1951 Lincolns.

When the Ford Motor Company introduced its completely new cars for model-year 1949, there were two distinct Lincoln series: the top-of-the-line Cosmopolitan and the Mercury-based small Lincolns. While the Cosmopolitans shared no sheet metal with any other Ford products, the small Lincolns were pure Mercury from the cowl back, except for distinctive Lincoln markings.

The front sheet metal, including the grille and the recessed headlamps, were like those used on the Cosmopolitans. Unlike the Cosmopolitans, which had a modern one-piece windshield, the small Lincolns (like their contemporaries from Ford and Mercury) stuck with a rather old-fashioned twin-paned windshield. The small Lincolns rode on a 121-inch wheelbase, compared to the Cosmopolitan's 125 inches and the Mercury's 118 inches.

Under the hood of the small Lincolns beat the same 336.7-cubic-inch/145-horsepower V-8 used in the Cosmopolitans. The Mercury used the 110- to 112-horsepower flathead V-8. The difference in weights between the Mercury and the small Lincoln was about 700 pounds. By stuffing a big engine in a smaller body, Lincoln had created what would later (in a decade or so) be called a "muscle car."

Performance-wise, the Lincoln could turn in top speeds of almost 97 miles per hour and 0-to-60 times of under 15 seconds, compared to 90 miles per hour and 17.5 seconds for Mercurys. Since Ford still did not have its own automatic transmission perfected, Hydra-Matic was available in 1949-1954 Lincolns.

1951 would be the last year for the small Lincolns, though. For many years hence, Lincoln would only use one basic body style. The 1951 small Lincoln was officially called the 1EL series. The previous years were called the 9EL for 1949 and the 0EL for 1950. In most years, there were only two basic versions, a six-passenger Club Coupe and a four-door Sports Sedan complete with rear "suicide" doors. Unlike the Mercury, there were no Lincoln convertibles in 1950 or 1951, nor were there station wagons in any year.

Price-wise, the 1EL fit between the Mercury and the Lincoln Cosmopolitan. For example, a 1951 1EL four-door listed for $2,553 compared to $3,182 for a comparable Cosmopolitan and $2,000 for a four-door Mercury. The small Lincolns were not popular cars. Less than 17,000 were made in 1951, compared to around 300,000 Mercurys. Part of the problem was the higher price for a car that looked too much like a lower-priced Mercury, especially from the rear.

The most desired small Lincoln has to be the Lido model. It was the equivalent of the Ford Crestliner and the Lincoln Capri, the cars Ford used to counter (rather unsuccessfully) the full line-up of General Motors hardtops in 1950-1951. In addition to a canvas or vinyl roof, the 1951 Lidos came with fender skirts, a custom interior and rocker panel chrome.

SPECIFICATIONS

Year	1951
Make	Lincoln
Model	L-74
Body style	four-door Sport Sedan
Base price	$2,553
Engine	L-head V-8
Bore x stroke	3.50 x 4.38 inches
CID	336.7
Compression ratio	7.0:1
Carburetor	Two-barrel
H.P.	154 @ 3600 rpm
Wheelbase	121 inches
Overall length	214.8 inches
Weight	4,130 pounds
Tires	8.00 x 15
OCPG Value	$16,000

Lincoln Sport Sedans "sported" three trim variations in 1951. This is the third type offered late in the model run. The small Lincolns were not popular cars. Less than 17,000 were made in the 1951 model run.

BUFFER FOR MAINE'S SEA-SWEPT COAST, FORTUNE ROCKS ENDURE YEAR AFTER YEAR.

*I*ndifferent to the passing years...

As nature's splendor endures throughout the years, so, too, does the satisfaction of Lincoln ownership.

The distinctive lines of the 1951 Lincolns will always be in style—for they are simple, sensible, and functional —not a designer's whim of the moment.

And all of the ingredients of the "Lincoln look"—from exterior paints to inside fabrics—have undergone the most rigorous tests for durability that the laboratory can devise.

Other proofs of Lincoln's indifference to the passing years come on the road. Test drivers, on roads and tracks deliberately built to strain the car to the utmost, have found that the Lincoln's velvet-smooth ride is designed to stay that way for more years than you will ever drive the car.

It's very likely true that the cost of a Lincoln is well under what you would expect.* See and drive the 1951 Lincoln or the magnificent Lincoln Cosmopolitan—and you will be convinced that *Nothing could be Finer*.

LINCOLN DIVISION • FORD MOTOR COMPANY

Standard equipment, accessories, and trim illustrated are subject to change without notice.

*PROOF OF ECONOMY. A 1951 Lincoln, with overdrive, out-performed all other cars in the famous 1951 Mobilgas Economy Run, winning the Sweepstakes Award.

Nothing could be finer — Lincoln for 1951

The small Lincolns were pure Mercury from the cowl back, except for distinctive Lincoln markings. "Indifferent to the passing years ... nothing could be finer ... Lincoln for 1951," boasted this advertisement.

191

$10,000 without air: the 1956 Continental Mark II

By Tom LaMarre

"The excitement it stirs in your heart when you see the Continental Mark II lies in the way it has dared to depart from the conventional, the obvious. And that's as we intended it. For in designing and building this distinguished motor car, we were thinking, especially, of those who admire the beauty of honest, simple lines ... and of those who most appreciate a car which has been so conscientiously crafted."

The original Continental was Edsel Ford's pet project. The new version was his son, William Clay Ford's, baby. It took three years to develop the Mark II. Management viewed the 13 proposals which were submitted by consultants and Ford's Special Products Division. Finally, management selected the design prepared by its own Special Products Division, whose design team included John Reinhart, Gordon Buehrig, Robert Thomas and Harley Cobb.

Because the Mark II was so long, Harley Cobb, chief engineer for the Special Products Division, designed a special frame to allow high seating with a low roof line. Unlike later Continentals, the stamped-in spare tire cover on the trunk lid actually covered the spare.

To market the Mark II, the Continental Division was created. Technically, the car is not a Lincoln Continental, but simply a Continental. The Mark II was introduced on October 6, 1955 at the Paris Auto Show. During the car's two-year production run, it was offered to the public in only one model, a two-door coupe. It sold for nearly $10,000 and came loaded with everything, but air conditioning, as standard equipment.

Two custom convertibles were built. One of them, built in Lincoln's own shop, was William Clay Ford's personal car. Derham built another one in 1957 for the Ford show fleet. After being shown at the Texas State Fair, it toured the country.

The Special Products Division also turned out plans for a retracting-hardtop convertible. A cut-away drawing marked "confidential" pictured a retractable 1956 Continental with its top in the up position. The details showed three power sources and five synchronized mechanism systems for the retractable hardtop.

The Continental Division was considering adding a convertible to the line for 1958, along with a four-door Berline. Instead, the Mark II was replaced by a less expensive and totally redesigned Mark III. The Continental Division itself was eliminated in 1961.

SPECIFICATIONS

Year	1956
Make	Continental
Model	Mark II
Body style	two-door hardtop
Base price	$9,966
Engine	OHV V-8
Bore x stroke	4.00 x 3.65 inches
CID	368
Compression ratio	10.0:1
Carburetor	Carter 4V
H.P.	300 @ 4800 rpm
Wheelbase	126 inches
Overall length	218.5 inches
Weight	4,825 pounds
Tires	8.20 x 15
OCPG Value	$29,000

The Continental Mark II was William Clay Ford's, baby. It took three years to develop the luxury car. Lincoln established a separate Continental Division to facilitate production and marketing. (Ford Archives photo)

Unlike later Continentals, the stamped-in spare tire cover on the trunk lid of the 1956 Continental Mark II actually covered the spare. The same basic car was marketed in 1957. (Tim Howley photo)

William Mitchell of Mitchell-Bentley Corporation of Ionia, Michigan leans on an unfinished Continental Mark II body. His firm helped with production of the expensive specialty cars. (Courtesy Joe Bortz)

A Mark II convertible was built in Lincoln's own shop as William Clay Ford's personal car. Derham built another in 1957 for the Ford show fleet. After being shown at the Texas State Fair, it toured the country. (Ford Archives)

Mark III embodied personal luxury

By Tom LaMarre

The Lincoln Continental Mark III caused a sensation when it was unveiled in February 1968. Two months later, it was in the showrooms.

Not everyone liked the new car. In an article in the October 1968 issue of *Popular Mechanics*, Bill Kilpatrick wrote, "Lincoln: Two cars here, the Mark III and the plain Continental. The Mark III was introduced earlier this year and is carried over as is, meaning a plush two-door coupe that strives to combine the concept of a personal luxury car with all-out luxury. Whether or not the idea is a success is in the eye of the beholder ... Ford knows I'm not nuts about the Mark III, but I think the regular Continental is handsome."

However, most people disagreed with the article and there was no question that the Mark III was a success. The car's Rolls-Royce type grille made it easily identifiable and the decorative spare tire cover on the trunk continued a styling element of the 1957 Mark II.

The confusing part was that Lincoln had already made a Mark III in 1958, when the designation was given to the top-of-the-line Lincoln after the Mark II was discontinued.

Ford didn't bother with an explanation in 1968. In fact, magazine ads for the Mark III had very little to say about the car. "The Continental Mark III" was printed in bold letters, surrounded by a lot of white space. The fine print called it, "The most authoritatively styled, decisively individual motorcar of this generation. From the Lincoln-Mercury Division of Ford Motor Company."

The thing that stood out in the ad was the car's highly reflective paint job. Television commercials emphasized the special dipping and finishing process used on Mark III bodies.

Like its rival the Cadillac Eldorado, Lincoln's personal/luxury model used the long hood/short deck styling formula. The Mark III was smaller than the standard Continental, but its styling was still formal, rather than sporty. Almost all Mark IIIs had vinyl tops.

An optional anti-skid braking device was introduced in the 1969 model-year. That's one of the minor changes the resurrected Mark III underwent. Its styling continued virtually the same through the 1971 model-year.

SPECIFICATIONS

Year	1968
Make	Lincoln Continental
Model	Mark III
Body style	two-door coupe
Base price	$6,585
Engine	OHV V-8
Bore x stroke	4.36 x 3.85 inches
CID	460
Compression ratio	10.5:1
Carburetor	4V
H.P.	365 @ 4600 rpm
Wheelbase	117.2 inches
Overall length	216 inches
Weight	4,475 pounds
Tires	9.15 x 15
OCPG Value	$12,000

The Lincoln Continental Mark III caused a sensation when unveiled in February 1968. It was introduced at midyear as a 1969 model. The front had a neo-classic look that proved very popular. (Ford Archives)

This February 13, 1968 news photo showed the Mark I, the mid-1950s Mark II and the new Mark III. Lincoln had first used the Mark III name in 1958 to identify the restyled Continental that followed the Mark II. (Ford Archives)

The decorative spare tire cover on the trunk of the 1969 Mark III continued a styling element introduced on the 1940 Lincoln Continental ... later called the Mark I ... and also used on the 1956 Continental Mark II. (Ford Archives)

"The most authoritatively styled, decisively individual motorcar of this generation," ads said. Like the rival Eldorado, Lincoln's personal/luxury car had long hood/short deck styling. (Henry Austin Clark, Jr.)

1949 Mercury woodie was different

By John A. Gunnell

Mercury's first line of all-new postwar cars hit the showrooms in April 1948. They represented a break with tradition. Instead of resembling fancy Fords, they were viewed as "baby Lincolns." The 1949 Mercury station wagon, however, was different. It was more Ford-like than the other cars in the series.

The station wagon had the new Mercury front end sheet metal. However, it had the same wood body used for Ford station wagons. All of Ford's station wagon bodies were constructed at a facility in Iron Mountain, Michigan. There was no wagon in the Lincoln line-up, so it made sense for Ford and Mercury to share the same wood body structure.

A change for 1949 was a switch to a two-door station wagon. The wagons actually featured all-steel construction. The maple and mahogany panels were genuine wood, but served mainly a decorative function. The body structure itself was metal.

Like other Mercurys, the station wagon had a 118-inch wheelbase. Its overall length of 213.53 inches was slightly greater than that of coupes, sedans and convertibles. The roof, floor pan, doors and quarter panels were metal. The liftgate and rear window were also framed in metal and a metal-covered spare tire was attached to the tailgate.

All cars were powered by the venerable flathead V-8, which was made more powerful for the Mercury in 1949. Previously, postwar Fords and Mercurys had shared the same engine, but this year the Mercury engine was stroked and gained 10 extra horsepower. Three-speed manual transmission with a steering column-mounted gearshifter was standard. Touch-O-Matic overdrive was available for slight extra cost, but an automatic transmission was not offered.

Wagons featured 3-2-3 seating with tan, green or red leather upholstery. A leather-grained headliner was used. On the outside, bright belt moldings were standard. Sliding quarter windows, automatic tailgate latches and a hinged taillamp that always stayed parallel to the ground were among other features of the station wagon.

Advertising copywriters described the new-generation Mercury as a "longer, lower, wider" car and bragged "its broad-beamed sturdiness is artfully combined with fleetness of line." The Mercury was smoother-riding, too, with coil springs replacing old-fashioned transverse leaf springs up front. Leaf springs remained at the rear. Axle ratios varied between 3.9:1 without overdrive and 4.27:1 with it.

Popularly received in the postwar sellers' market, sales of all 1949 Mercurys leaped to 301,319 units versus about 80,000 the previous season. However, the output of wagons increased only modestly in comparison to overall totals and 8,044 woodies were made.

SPECIFICATIONS

Year	1949
Make	Mercury
Model	9CM "Woodie"
Body style	two-door station wagon
Base price	$2,716
Engine	L-head V-8
Bore x stroke	3.19 x 4 inches
CID	255.4
Compression ratio	6.8:1 (Manual)
Carburetor	Holley 2V
H.P.	110 @ 3600 rpm
Wheelbase	118 inches
Overall length	213.53 inches
Weight	3,626 pounds
Tires	7.10 x 15
OCPG Value	$20,000

The 1949 Mercury station wagon was more Ford-like than the other cars in the series. One change was the two-door wagon body. The body structure itself was actually metal. The production of 1949 wagons increased modestly to 8,044.

Restorable "woodies" are still around if you look hard enough. The Mercury station wagons featured 3-2-3 seating with tan, green or red leather upholstery. A leather-grained headliner was used.

"There's no match for MERCURY style!"

THAT'S WHAT OWNERS SAY ABOUT THE HANDSOME NEW 1949 MERCURY!

It's long! It's low! It's really massive! No wonder owners say there isn't a smoother, smarter, more handsomely designed car than the 1949 MERCURY! And they're right! There isn't!

Make your next car

MERCURY

YES, the 1949 Mercury is the most beautiful value today – in more ways than one!

For everything in it has been *road-proven* by thousands of owners over millions of miles!

You get a new 8-cylinder, V-type engine with surprising *economy!* Front coil *springing!* A truly restful "comfort-zone" *ride!* Easier *steering!* "Super-safety" *brakes!* Increased all-round *visibility!* Plus the luxury of foam rubber-cushioned *seats!*

See it! Drive it–and you'll say: *"It's Mercury for me!"*

MERCURY DIVISION OF FORD MOTOR COMPANY

White side-wall tires and rear wheel shields optional at extra cost

"There's no match for Mercury style!" boasted this advertisement, which featured every basic model except the station wagon. "It's long! It's low! It's really massive!" sold cars then, but wouldn't nowadays.

The last of Mercury's Marauders:
1969-1970

By Tom LaMarre

"I can't make up my mind whether the Mercury is being upgraded to look like the Lincoln or the Lincoln is being downgraded to look like the Mercury," Bill Kilpatrick wrote in *Popular Mechanics*. "In any event, there's a definite resemblance, particularly up front."

The Marauder name had a rich performance heritage dating back to the 1963-1/2 models successfully campaigned by Parnelli Jones. The 1969 version came in two series ... the regular Marauder and the high-performance, low-production Marauder X-100. There was only one X-100 body style, a fastback hardtop coupe weighing 4,009 pounds and having a factory price of $4,074.

The Marquis-style front end with concealed headlamps was classic Lincoln all the way. However, the Marauder's rear end styling was unique. A tunneled rear window added a sporty touch and the quarter panels were decorated with simulated air scoops. Individual rectangular taillights were recessed in the back panel. There were three taillights and one back-up light on each side with Marauder script separating them.

The X-100 came with many standard features not shared by the plain-vanilla Marauder. They included styled aluminum wheels and fiberglass-belted H70 x 15 tires. The interior was trimmed in vinyl and cloth. Among the X-100 options were high-performance and power-transfer axles and a competition handling package.

The X-100's 429-cubic-inch engine with 360 horsepower was coupled to a SelectShift automatic transmission. "A real marcher," is how *Popular Mechanics* described it.

Despite 1969 model-year production of only 5,635 cars, the Marauder X-100 was back in 1970 with only minor changes. Instead of a Mercury crest, "Marauder" was spelled out in block letters above the grille. Vertical chrome strips were added to the grille and parking lights. Matte black paint on the rear section continued to be a popular option.

An unusual factory photo of a 1970 Marauder X-100 indicates that Lincoln-Mercury was considering toning down its luxury-performance model. The prototype in the photo lacked fender skirts and wheel opening trim and the use of a vinyl top eliminated the Marauder's semi-fastback appearance. The twin upper body stripes were also missing. However, the prototype did have high-back bucket seats (with no separate headrests), super-fat tires that looked as if they were borrowed from road-building equipment, and 1968 Mercury wheelcovers.

Only 2,646 of the 1970 Marauder X-100s were built. The regular Marauder didn't fare much better. Its production was 3,397 cars, which were the last Marauders.

SPECIFICATIONS

Year	1969
Make	Mercury
Model	Marauder X-100
Body style	two-door hardtop
Base price	$4,074
Engine	OHV V-8
Bore x stroke	4.36 x 3.59 inches
CID	429
Compression ratio	10.5:1
Carburetor	4V
H.P.	360 @ 4600 rpm
Wheelbase	121 inches
Overall length	219.1 inches
Weight	4,074 pounds
Tires	H70-15
OCPG Value	$8,000

SPECIFICATIONS

Year	1970
Make	Mercury
Model	Marauder X-100
Body style	two-door hardtop

Base price	$3,873
Engine	OHV V-8
Bore x stroke	4.36 x 3.59 inches
CID	429
Compression ratio	10.5:1
Carburetor	4V
H.P.	360 @ 4600 rpm
Wheelbase	121 inches
Overall length	219.1 inches
Weight	4,128 pounds
Tires	H78-15
OCPG Value	$8,000

The Marauder name dated to 1963-1/2 models raced by Parnelli Jones. The 1969 version came as a Marauder or low-production Marauder X-100. The X-100 had styled aluminum wheels and fiberglass-belted H70 x 15 tires. (Phil Hall)

Marquis-style front with concealed headlamps was classic Lincoln all the way. The rear end styling was unique. The X-100's 429-cubic-inch engine with 360 horsepower was coupled to a SelectShift automatic transmission. (Phil Hall)

Oldsmobile's B-44

By Robert C. Ackerson

Oldsmobile began production of its 1942 models on September 9, 1941 and less than six months later, on February 3, 1942, the company finished putting the final touches on the last of the 1942 models to be manufactured.

Even before the Japanese attack on Pearl Harbor, the American automobile industry was deeply involved in the rearming of America. In mid-1941, it was the government ... not the consumers ... who told Detroit how many cars it would be turning out.

As did the rest of the industry, Oldsmobile combined patriotism with a traditional sales pitch in promoting its 1942 models. In Oldsmobile's case, this extended to describing the 1942 in military jargon. "To serve the vital needs of today's America," read one ad in *The Saturday Evening Post* (October 11, 1941). "The nation's oldest motor car builder contributes the B-44."

Oldsmobile quickly explained that the B-44 designation represented a car that was "even better looking, even better lasting, even better built than any Oldsmobile in 44 years." But, with the new Oldsmobile sharing its space with a P-39 Aircobra and an army artillery piece, the B-44 name was definitely in tune with the times.

Obviously not concerned about the possibility of losing a few sales to pacifists, Oldsmobile touted the B-44 as as car "stamina-styled" in a bold, new military mode ... with new "fuselage" fenders, a new "drednaught" frame, and an engine of even greater "fire power." In calmer times, these terms would seem a bit extreme, but this was 1941 and, for millions of Americans, "war talk" was part of their every day conversation.

Oldsmobile, like other companies, had to work hard to produce cars that maintained previous quality and performance standards levels in the context of War Production Board restrictions on the use of strategic materials such as zinc, aluminum, nickel and chromium. Common to all cars was an almost complete elimination of die castings and stainless steel. At the same time, there was a widespread adoption of alloy iron pistons and cylinder heads in place of aluminum. For example, both Oldsmobile's six- and eight-cylinder engines were fitted with cast iron "Armasteel" pistons in place of the aluminum type used in 1941. In order to hold down the use of zinc, very few die castings were used for the B-44 models.

Yet, not every aspect, by no means, of the B-44 Oldsmobile's character was somber. Up front was a "double-duty" bumper structure that swept away the old barriers that for years had been maintained between the grille and the bumpers. The use of both broad and narrow horizontal bars spaced across the Oldsmobile's front end wasn't an artistic triumph, but it did serve to bridge the gap that would later be seen as existing between prewar and postwar automobile design.

Early B-44s carried a full complement of bumper-grille units, but as 1941 came to an end, this symbol of the good old times disappeared. At first, the lower grille section was abandoned to body-color-painted elements. By December 31, 1941 only bumpers were allowed to be chrome plated. The remaining portions of the Oldsmobile grille were then painted dark brown and tan with white striping.

All Oldsmobiles were equipped with front fenders that faded away into the front door panels, like many General Motors cars. The Oldsmobile's fenders weren't as dramatically styled as those on Buick's convertible and sedanette models in its 50 and 70 car-line, which swept all the way back to join the rear fenders.

Oldsmobile offered its 100-horsepower six-cylinder 238.1-cubic-inch engine in three series: Special 66, Special 68 and Dynamic Cruiser 76. Two other series, the Dynamic Cruiser 78 and Custom Cruiser 98 were powered by a straight eight displacing 257.1 cubic inches and developing 110 horsepower at 3500 rpm. Both of these power outputs were unchanged from 1941, but there were changes made for 1942. The compression ratio of the six was raised from 6.1:1 to 6.5:1, while that of the eight moved up to 6.5:1 from 6.3:1 Both engines also had redesigned combustion chambers and sintered main bearings. On the eights, a stronger crankshaft and connecting rods were also used. Found on both engines was a groove running down the top of the cylinder head, which was intended to drain water away from the spark plugs.

Other improvements found on the 1942 models were a stiffer frame, an additional front cross member (to give increased stability to the double-duty bumper) and an improved battery cover with a dust shield removable without tools. Oldsmobile convertibles used two electric motors in place of vacuum cylinders to operate their tops. Braking performance was enhanced by the adoption of quarter-inch wider front brake linings.

Total Oldsmobile production for the abbreviated 1942 model year was just 67,999 units. Contrary to what many people feared, these were durable vehicles that served their owners well, not only during the war years, but for many peacetime months before they gave way to a new generation of Futuramic Oldsmobiles.

SPECIFICATIONS

Year 1942
Make .... Oldsmobile
Model .... Special 66 (B-44)
Body style .... two-door Club Coupe
Base price $1,050
Engine .... L-head; six-cylinder
Bore x stroke .... 3-1/2 x 4-1/8 inches
CID 238.1
Compression ratio 6.5:1
Carburetor .... Single downdraft
H.P. .... 100 @ 3400 rpm
Wheelbase .... 110 inches
Overall length .... 204 inches
Weight .... 3,205 pounds
Tires .... 16 x 6.00
OCPG Value $8,000

Up front was a "double-duty" bumper structure that swept away the old barriers that had been maintained between the grille and the bumpers. The convertible was a particularly rare body style. (Oldsmobile photo)

Obviously not concerned about offending pacifists, Oldsmobile touted the B-44 as a car "stamina-styled" in a bold, new military mode. This advertisement stressed the "quality-built to last" to buyers on the home front.

The 1942 Oldsmobile 98 sedan was apt to draw a crowd. Oldsmobile began production of 1942 models on September 9, 1941 and ended the model run early, on February 3, 1942, due to the outbreak of World War II. (Oldsmobile photo)

All Oldsmobiles were equipped with front fenders that faded away into the front door panels. This 1942 Oldsmobile Special 66 B-44 Club Coupe sold for just over $1,000 when it was new.

1958 Oldsmobile: mid-price sales leader

By Robert C. Ackerson

Modern day critics of the automobile tirelessly render attack after attack upon the 1958 Oldsmobile as the epitome of all that was inept or in bad taste about the American automobile in the late 1950s. The main reason for their brutal treatment of the Oldsmobile is, of course, its lavish use of chrome trim.

Even in 1958, the Oldsmobile's styling was controversial. An article in the March 1958 issue of *Motor Life* by Stan Mott and Robert Cumberford noted that "the most shocking thing about the car is the confusion it engenders in the mind of the observer. There is too much going on for easy comprehension. The overall impression is that this is a very old and ordinary design tricked up by a styling staff."

Oldsmobile made no apologies for its styling format, calling it the "mobile look" and noting that it was "unmistakably a 1958 model ... every new line and feature is styled in finer taste."

Oldsmobile did, however, provide its sales personnel with plenty of verbal put-downs of the competition's styling. For example, Mercury and Edsel were described as having "various 'unusual' styling features ... (that) have produced mixed reviews in some quarters."

Oldsmobile also made the most of the Edsel's less-than-record-setting sales pace. "There is no mistaking the Edsel," admitted Oldsmobile. "It is unusual enough in every view to be readily recognized anywhere. But this unusual design had not yet been a sound basis for popularity. Then, too, the Edsel has not yet proved itself in the hands of the driving public. And there is also some uncertainty about Edsel's future re-sale value. These are facts that a buyer should consider." On the other hand, the Lansing, Michigan automaker described its own "Mobile Look" as "impressive and appealing."

Aside from subjective judgments of the Oldsmobile's styling relative to the competition's, a look at the medium-price car market in 1958 suggests that many new-car buyers liked what they saw in an Oldsmobile. That was not a particularly good year for the automobile industry as a whole, since production fell to its lowest level since 1948 when 3,910,700 cars had been assembled.

In contrast to the record output of 7,942,200 cars in the 1955 model-year and the respectable level of 6,115,400 reached in 1957, only 4,244,000 cars were produced to 1958 specifications. However, as far as Oldsmobile was concerned, its share of the medium-priced car market held up relatively well:

Model Year Production Comparison of U.S. Cars

Marque	1956	1957	1958	Change 1957-58
Oldsmobile	485,458	384,390	296,375	-22.9%
Pontiac	405,429	333,473	216,982	-35%
Buick	572,024	405,098	241,908	-40.3%
Mercury	327,943	286,163	133,271	-53.5%
Edsel	NA	NA	NA	NA
Chrysler	128,322	122,273	63,671	-48.4%
Dodge	233,686	281,359	133,953	-52.4%
DeSoto	111,422	133,685	49,445	-63.1%

Although the auto industry toned down talk of performance and power in 1958, Oldsmobile made much of its leadership role in modern overhead valve V-8 engine design. "More than 3.5 million Rocket engines have been built and tested on the rugged proving ground of customer ownership and loyalty," boasted Oldsmobile. In contrast, the new engines from Ford were depicted as "yet to prove themselves." Oldsmobile also noted that it "had actually reduced horsepower in 1958 for the sake of economy on Dynamic models. But, for the "performance-minded, there is J-2, first offered by Oldsmobile in 1957 and since copied by Ford and Mercury, but not Edsel."

In contrast to *Motor Life's* caustic comments, it was *Hot Rod* magazine's Ray Brock who made the most important remark about the 1958 Oldsmobile: "The leader in (the) Detroit chrome race with 44 pounds of glitter, Olds is also the sales leader in the medium-price class."

Sometimes bright cars do finish first.

SPECIFICATIONS

Year...1958
Make..Oldsmobile
Model...Dynamic 88
Body style ....two-door "Holiday" hardtop
Base price $2,893
Engine...OHV V-8
Bore x stroke ....4.00 x 3.69 inches
CID..371
Compression ratio 10.0:1
Carburetor 2V
H.P. ....265 @ 4400 rpm
Wheelbase...122.5 inches
Overall length ....208.2 inches
Weight...4,112 pounds
Tires..8.50 x 14
OCPG Value $14,000

With a continental tire, the Oldsmobile was even bigger and brighter. Hot Rod magazine said, "The leader in (the) Detroit chrome race with 44 pounds of glitter, Olds is also the sales leader in the medium-price class."

OLDSmobility

A new, 'free-and-easy' way
of going places and doing things in this mobile era!

Scan this sparkling span of lighthearted beauty—Oldsmobile for '58! Nearly eighteen feet of sheer excitement . . . captured in a tasteful new mobile look. There's new magic in its motion, too, with the "big-economy-news" Rocket Engine . . . and exclusive New-Matic Ride*, Oldsmobile's *true* air suspension. So come take a test ride at your dealer's soon. See how Olds for '58 makes real driving pleasure the rule on any road!

OLDSMOBILE DIVISION. GENERAL MOTORS CORPORATION

*Optional at extra cost.

OLDSMOBILE

This 98 Holiday Coupe parked near the Golden Gate Bridge had a little bit fancier trim than the 88. Oldsmobile held up its share of the medium-priced car market relatively well during 1958. The company lost a lower percentage of business than other competing automakers.

209

"Oldsmobility" was a term used in 1958 to pitch ride, performance and convenience. *"More than 3.5 million Rocket engines have been built and tested on the rugged proving ground of customer ownership and loyalty,"* said Olds.

Luxury in a large package: 1973 Oldsmobile 98

By John A. Gunnell

Oldsmobile Division built 939,630 car in the 1973 model-year. About half of them were "small" cars, such as the new Omega and the totally redesigned mid-sized Cutlass. The balance of production went to big cars, such as the Delta 88, the Toronado and the Ninety-Eight. The latter model, aimed at the lower fringe of the luxury car market, generated 133,141 assemblies.

Models available in the Ninety-Eight line-up ranged from the $4,796 coupe to the $5,417 Regency four-door sedan. In the middle of the pricing ladder was one of the most attractive models, the Ninety-Eight Luxury two-door hard-top. It combined a dash of elegance with a touch of sportiness.

Standard equipment in all Ninety-Eights included deluxe armrests; dual ashtrays; power brakes with front discs; interior hood latch; lamp package; molding package; windshield radio antenna; power seats; power steering; deluxe steering wheel; spare tire cover; and Turbo-Hydramatic transmission. The upholstery choices for these cars included vinyl, cloth and leather combinations.

Styling of the full-size 1973s dated back to 1971. That's when Oldsmobile first introduced the basic body that was used all the way through 1977. At that point, a corporate down-sizing program took effect. General styling features of the 1971-1977 big-cars included body side surfaces with an increased curvature, flush-mounted curved glass windows, slim front roof pillars and cantilever styled roofs strengthened by the addition of a full inner panel.

For 1973, the split front grille sported an egg-crate type insert. It was divided by a body-color panel bearing the Oldsmobile crest. A shelf-like chrome bumper ran across the grille. Upside-down trapezoid-shaped parking lamp lenses appeared directly under the bumper in a line below the dual horizontal headlamps. Side trim consisted of a slim rub molding running down the center of the body. On the Luxury Coupe, the molding began near the trailing edge of the front fenders and ran across the doors and the rear quarter panels.

Rear end styling was highlighted by a bright beauty panel running across the body just below the lip of the deck lid. It contained the back-up lamps, spaced out close to the center, and red rectangular reflectors near each end. The vertical taillamps had thick bright housings with narrow chrome center dividers and sat at the end of each fender. The upper portion of the fenders was shaped almost like a shallow tailfin and looked like a tribute to a classic 1950s styling motif.

A prominent feature line (or skeg) ran across the lower sides of the body. It angled down at the bottom of each wheel opening, giving them a sculpted, bulging contour. This bulge was also incorporated into the rear fender skirts, which curved outwards along their bottom half to match the body contours. Wheelcovers were of a simple, but elegant flat-disk design with the Oldsmobile logo in the center. The double-band whitewalls used in 1971-1972 were replaced with tires having a single narrow band of white.

According to surveys of original 1973 Oldsmobile owners, they were great cars. Questionnaires determined that they gave better than average service and would travel many trouble-free miles with proper car care. The Oldsmobile V-8 engine, fuel system and brakes were all very highly regarded by owners. They reported that far less than average repairs were needed in those areas. Better than average performance was also delivered by electrical components, the exhaust system and the Turbo-Hydramatic transmission.

Offering about average durability and service were Oldsmobile's body work, driveline parts and suspension components. The body was somewhat rust and corrosion-prone. These were also heavy cars, which affected the longevity of some steering and suspension parts. Owners reported above-average repairs to these, but overall the marque gave excellent reliability.

Naturally, the 455-cubic-inch V-8 was a bit of a gas guzzler. However, its 8.5:1 compression cylinder heads were designed for operation with unleaded fuels. This is a definite plus for the collector who drives his car regularly. Restorers may also find that the bumpers on these cars are somewhat prone to rust even more rapidly than the rest of the body.

Oldsmobile built a total of 24,452 Luxury Coupes. Many of these fine cars survive. They are perfect for use as "everyday" collector cars that can be driven regularly, while providing the potential to become collectible in the future. Today, though, they're available and affordable.

SPECIFICATIONS

Year 1973
Make .... Oldsmobile
Model .... Ninety-Eight
Body style .... Luxury Coupe
Base price $5,070
Engine .... OHV V-8
Bore x stroke .... 4.125 x 4.250 inches
CID 455
Compression ratio 8.5:1
Carburetor .... Rochester 4V
H.P. .... 275 @ 3600 rpm
Wheelbase .... 127 inches
Overall length .... 230.2 inches
Weight .... 4,601 pounds
Tires .... J78 x 15
OCPG Value $4,700

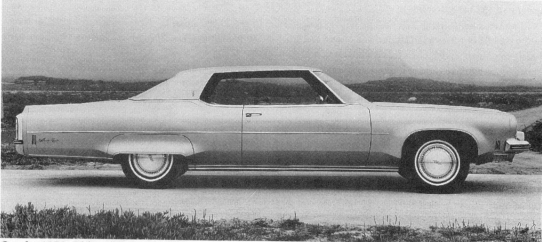

On the 1973 Oldsmobile Luxury Coupe the side molding began at the trailing edge of the front fenders and ran over the doors and rear quarter panels. This model came with a host of convenience and trim features. (Oldsmobile photo)

The front end had a two-section egg-crate grille with body-color center divider. Oldsmobile built a total of 24,452 Luxury Coupes. Many survive. They make excellent "everyday-use" collector cars. (Oldsmobile photo)

From the rear you can see the prominent skeg running down the lower sides of the body. It angles down at the bottom of each wheel opening, giving it a sculpted, bulging contour. (Oldsmobile photo)

The 1971 Oldsmobile 98 Coupe. Styling of the full-size 1973s dated back to 1971. That's when Oldsmobile first introduced the basic body that was used all the way through 1976. (Oldsmobile photo)

New management pushed 1953 Packard Patrician

By Gerald Perschbacher

"Packard has new top management and is out for sales with vigor," wrote the venerable Floyd Clymer in 1953. "In the luxury field, Packard is out to regain the prestige it lost when it brought out a less expensive line during the depression years."

Clymer spoke to Packard owners about the new 1953. They commented that the car was roomy and comfortable, full of trouble-free miles and had a certain identity. ("When you see a Packard on the highway, you don't have to guess what kind of car it is.") Should a Packard need servicing, one owner said, "We found no dealer as obliging as Packard's." Clymer's statements were based on replies from 1,000 Packard owners. He took the company's slogan at its word: "Ask the man who owns one."

The company sales force trumpeted the 1953 Packard with a loud horn, a sign of the new sales strategy pursued by James Nance, Hotpoint "hot shot" and new president of the ailing Packard enterprise. "Now ... look to Packard for '53," said its literature. "Engineered to outperform ... built to outlast them all. Designed to express your own good taste, with Packard advanced contour styling that sets the new trend in automobile design."

True, the contour style was born on the 1951 models. But, also true was that the styling had been advanced with minor touches here and there, compared to the Cadillac's stodgy, bulky front. In a Packard, you knew where your borders lay. The style forecast the trend in the mid-1950s, when General Motors styling featured the high fenders.

The 1953 Packard reached 60 miles per hour in about 18 seconds. Most owners (92 percent) used premium gas and got as high as 15 miles per gallon on their Packards. Visibility was rated as great; 98 percent liked the front view, 92 percent favored the rear. The ride pleased 90 percent of the owners; the engine's power pleased 80 percent; but only 55 percent had no problems with the engine's sound inside the car.

Why did people buy the 1953 Patrician or any Packard back then? Previous Packard ownership rated first, styling was second, and third was quality received for the money. Driving ease, riding comfort, and the trade-in on the old car is what brought other buyers to Packard's door.

The Patrician was top banana in Packard handling, pleasing 93 percent, who rated it as excellent, compared to the Packard Clipper Deluxe, which made only 72 percent of its owners ride on "Cloud Nine." A solid 76 percent of the Patrician owners claimed they would go back to Packard for their next car.

There were complaints of course. The highest were for leaks in the beautiful body (each model rated at 13 percent), while poor paint and chrome displeased seven percent. "Defense chrome" ... a result of the Korean War effort ... was a bane due to its short life, caused by the lack of base metal. Six percent said that acceleration was poor and the body was noisy. In total, only 45 percent of the owners of higher-priced Packards complained.

Packard's sales push brought it close to 25 percent of overall luxury car production, a mark much higher than the three percent under the prior management. "Packard definitely is on the comeback trail toward strengthening its prestige position. With its new models, owner satisfaction is higher than at any time in the firm's long history," Clymer said. "If you are a quality-car prospect, it will pay you to look over the new Packard line," he concluded.

SPECIFICATIONS

Year	1953
Make	Packard
Model	Patrician
Body style	four-door sedan
Base price	$3,735
Engine	Inline 8
Bore x stroke	3.50 x 3.25 inches
CID	327
Compression ratio	8.0:1
Carburetor	Carter WGD 2V
H.P.	180 @ 4000 rpm
Wheelbase	127 inches
Overall length	218-5/32 inches
Weight	4,335 pounds
Tires	8.00 x 15
OCPG Value	$16,000

BUILT FOR THOSE
WHO WANT THE FINEST...

America's Most Advanced New Car!

PACKARD

DISTINCTION speaks in many ways—but never more eloquently than when you drive *America's new choice in fine cars*—Packard! Today's brilliant new Packard offers you unmatched performance and luxury in your choice of eight magnificent models.

HERE YESTERDAY'S TRADITIONS of craftsmanship join tomorrow's advanced engineering to bring you everything you have desired in a car, and probably much you never dreamed possible—the incredible smoothness of the Packard ride, for example . . . the effortless ease of Packard power steering . . . and the hush of the mighty Packard Thunderbolt-8 Engine—when "loafing" at sixty!

IT IS SIGNIFICANT that Packard, which owes allegiance to no other brand name, builds the one fine car expressing real *individuality*—its own and yours.

NOW . . . ASK THE MAN WHO OWNS ONE

See The Exciting New Packard CLIPPER For Big-Car Value At Medium-Car Cost

The Patrician was the "top banana" Packard in handling, pleasing 93 percent of buyers who rated it excellent on the road. Packard owners told Floyd Clymer the new 1953 was roomy and comfortable, full of trouble-free miles and had a certain identity.

New Packard CLIPPER

New! Big-Car Value At Medium-Car Cost!

America's Oldest Maker of Fine Cars Introduces
An Entirely New Line—The Packard CLIPPERS!

EVERY INCH a Packard in fine engineering and craftsmanship, the new *Packard CLIPPER* is designed for those who want outstanding quality, yet wish to spend only a few hundred dollars more than for a car in the low-price field.

HERE YOU GET advanced contour styling—a Packard first that's setting the new trend in motorcar smartness. You enjoy the flashing performance of the brilliant new Packard Thunderbolt-8 Engine. What's more, you get more room, better visibility, cushion-mounted safety bodies and direct-acting shock absorbers—all of which combine to give the beautiful new *Packard CLIPPER* the matchless smoothness, luxurious comfort and quiet of the famous Packard ride!

COME, SEE AND DRIVE the new *Packard CLIPPER*—now on display in a wide range of models. A true product of Packard engineering, it is today's BIG NEWS in the medium-price field!

For America's new choice in fine cars...see the new PACKARD—today's most advanced motorcar

The Patrician was tops in handling and 93 percent rated it excellent. It compared favorably to the Clipper Deluxe shown here, which put only 72 percent of buyers on "Cloud Nine."

Zig-zag Packard

By Jane A. Schuman

In 1954, Packard produced a scant 31,291 cars. People wondered if the company was going under. When Studebaker and Packard merged, new management assured the public that it would "keep the Packard tradition unbroken."

The new group brought together a group of skilled specialists to produce cars of distinction and individuality. The cars produced by the Studebaker-Packard team, they said, would be able to measure up to the demands of the most critical buyers and earn "owner approval when anyone decides to ask the man who owns one."

Was this statement merely corporate hype? Not if you judge the new management's initial work. Body styling was revamped to show a new design. The 1955 Packards featured a wraparound windshield. There was also a wide range of eye-catching two-tone and tri-color paint combinations.

A new Custom Clipper came in two models: four-door sedan and the Constellation two-door hardtop. The Constellation was advertised as, "The brightest star in the hardtop galaxy."

The Constellation was available with flashy two-tone paint schemes. A double color-sweep side molding treatment was a standard feature of this model. It had a "zig-zag" appearance and came in color schemes such as black and yellow.

In this particular case, black paint swept down the sides of the front fenders and part of the front doors up to a level even with the top of the wheel opening's arc. Black paint also swept along the upper rear quarters in a slim band, as well as over the entire rear end of the car. Black was used on the roof, too. The rest of the front doors, upper front fenders and hood would be yellow.

The Constellation's two-toning looked quite novel. It certainly provided this model in the Clipper Custom line with a touch of the distinction and individuality that Packard's new management had promised.

Packard sales for 1955 came to 52,244 cars. Of these, only 6,672 were model 5567 Clipper Constellations. Overall business for the marque was a vast improvement over the previous year, but the Studebaker connection did not work out as a long-term solution to the company's problems. In 1956, the last of the traditional Packards came off the assembly line. For 1957 and 1958, the Packard became a Studebaker clone with some extra features.

SPECIFICATIONS

Year 1955
Make .... Packard
Model .... Custom Clipper
Body style .... Constellation two-door hardtop
Base price $3,076
Engine .... OHV V-8
Bore x stroke .... 4.00 x 3.50 inches
CID 352
Compression ratio 8.5:1
Carburetor .... Carter WCFB 4V
H.P. .... 245 @ 4600 rpm
Wheelbase .... 122 inches
Overall length .... 215.3 inches
Weight .... 3,865 pounds
Tires .... 7.60 x 15
OCPG Value $16,000

This golfer probably liked the Clipper Custom Constellation hardtop's large trunk. The 1955 Packards featured a wraparound windshield. There was also a wide range of eye-catching two-tone and tri-color paint combinations.

Packard sales for 1955 came to 52,244 cars. Of these, only 6,672 were Clipper Constellation hardtops. This example was brought back to showroom condition by Batista Restorations of Southern California.

For those who desire individuality

CUSTOM CONSTELLATION
245 HORSEPOWER

The medium-priced car that makes it smart to be different

THE 1955 CLIPPER offers a distinctiveness . . . not confined by convention . . . unmatched by the commonplace. Here is a *smartness* that is *new* in its field . . . from sparkling exteriors to color-coordinated interiors. Two great V-8 engines, 225 and 245 horsepower, are the most powerful of all in this price class. There is Twin Ultramatic — the industry's most advanced automatic transmission — actually two in one . . . offering the driver a finger-tip selection of super-smooth starts or jet-like getaways.

Small wonder people with a desire for individuality so warmly welcome this car which sets both itself and its owner apart from the crowd. There's nothing uncertain in their approval . . . they are trying and buying the 1955 Clipper in unprecedented numbers. *One visit to your Packard dealer* will show you the reason for this tremendous reception and convince you the 1955 Clipper is the most individually distinctive car in the medium-price field.
PACKARD DIVISION • STUDEBAKER-PACKARD CORPORATION

The 1955 *Clipper*

Built by Packard Craftsmen

Advertised "for those who desire individuality," the Custom Constellation was available with flashy two-tone paint schemes. A double color-sweep side molding treatment was a standard feature of this model.

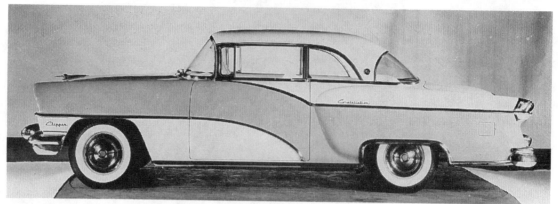

Seen in profile, the Constellation's two-toning looked novel. It provided this model from the Clipper Custom line with a touch of the distinction and individuality.

The new Custom Clipper came in two models: four-door sedan and the Constellation two-door hardtop. The Constellation was advertised as, "The brightest star in the hardtop galaxy."

1957 Packard Clipper Country Sedan

By Bill Siuru

Today, turbocharging is a way many automakers get peak performance from small displacement engines. Back in the 1950s, a couple of marques could be ordered with supercharged engines. These included the 1954-1955 Kaiser Manhattans, 1957 Fords equipped with the Thunderbird Special supercharged V-8 and the 1957-1958 Studebaker Golden Hawks.

However, most people do not realize that the 1957 Packard claims the honor of being the supercharged car of that era with the largest production volume. That is because all 4,809 Packards in the 1957 model year came with a supercharged engine. This included 869 station wagons.

These station wagons represented the only serious production of supercharged station wagons by any American automaker until Studebaker offered R-2 Avanti supercharged engines as an option for 1963-1964. Just for the record, some 4,325 Kaiser Manhattans were sold in the United States in 1954 and all of them were also supercharged.

Diehard Packard fans will tell you that the 1957 models were not really Packards, but badge-engineered Studebakers. This was the result of the Studebaker/Packard merger of 1954. All 1957 Packards were Clippers and there were only two models to choose from, the regular Town Sedan and the Country Sedan. The latter was actually a station wagon.

Both cars were really Studebaker Presidents with different finned rear fender treatments, more elaborate trim, pre-"Packard-baker" inspired paint schemes and, of course, Packard emblems. The Country Sedan was based on the Studebaker Broadmoor six-passenger station wagon.

Mechanically, the main difference between the 1957 Studebaker and the 1957 Packard was under the hood. While both used the Studebaker 289-cubic-inch V-8, the Packard version gained an additional 50 horsepower by using a McCulloch supercharger. It clicked in at 3000 rpm and gave the engine a 275 horsepower rating. This was the same amount of horsepower produced by the 352-cubic-inch Packard-built V-8 used in the 1956 Clippers.

The same supercharged engine was also used in 1957 and 1958 Studebaker Golden Hawks. However, it was not available in the balance of the Studebaker model line-up.

Contemporary motor magazine tests showed that the 1957 Clippers could turn in 0 to 60 miles per hour times of under 11 seconds. Because of the "compression boost" that came with the blower, the Packard engine had a 7.8:1 compression ratio, compared to the (unsupercharged) Studebaker V-8's 8.3:1. And, while the four-barrel carburetor was used to get 225 horsepower from the unblown engine, a two-barrel was used on the Packard power plant. Besides the supercharged engine, the Clipper also came with Flight-O-Matic automatic transmission as standard equipment.

The Country Sedan rode on the same 116.6-inch wheelbase as all the Studebaker wagons, but was a couple of inches longer in overall length. Price-wise, the Country Sedan listed for $700 more than the Broadmoor.

The 1957 Packard Clipper Country Sedan is an interesting vehicle, not only because it represents the only early mass-produced supercharged station wagon made in America, but also because it was the first Packard station wagon since 1950.

While the Packard name would linger on one more year, 1958 Packard production included a mere 159 wagons. They all used the standard Studebaker 289-cubic-inch V-8 with 225 horsepower. However, the unique 1958 Packard Hawk continued to use the supercharged engine.

SPECIFICATIONS

Year 1957
Make .... Packard
Model .... Broadmoor Country Sedan
Body style... four-door station wagon
Base price.. $3,384
Engine .... OHV V-8
Bore x stroke... 3.56 x 3.63 inches
CID 289
Compression ratio... 7.8:1
Carburetor.. Stromberg 2V
H.P. .... 275 @ 4800 rpm
Wheelbase .... 116.5 inches
Overall length .... 206.2 inches
Weight .... 3,555 pounds
Tires .... 8.00 x 14
OCPG Value $8,400

The Country Sedan rode on the same 116.6-inch wheelbase as all the Studebaker wagons, but was a couple of inches longer in overall length. All 4,809 Packards in the 1957 model year came with a supercharged engine.

All 1957 Packards were Clippers. There were two models to choose from, Town Sedan and Country Sedan station wagon. Model-year production included 869 wagons. The Clippers could turn in 0 to 60 times of under 11 seconds.

Packard's 1958 production included a mere 159 wagons. They grew higher fins and used the standard Studebaker 289-cubic-inch V-8 with 225 horsepower. However, the 1958 Packard Hawk continued to use the supercharged engine.

The 1950 Plymouth

By Robert C. Ackerson

Certain years of the automotive modern era stand out as pivotal points around which key styling, performance and technical themes were developed. For Plymouth, 1949 was such a year. So, no one expected any changes of great significance to be found in the 1950 Plymouths. However, a close comparison of the 1949 and 1950 models indicates that they are set apart in numerous ways.

At the front in 1950 was a grille with a single center bar. Both the front and rear fenders of 1949, with their dignified but dated horizontal fluting, gave way to units with smooth surfaces in 1950. Unlike 1949, when it was integrated into the grille design, the 1950 model's emblem was positioned above the enlarged "Plymouth" lettering on the hood. Also, the stylized sailing ship hood ornament was moved further forward on the hood.

Also easy to detect was the 1950 Plymouth's much larger and curved rear window. Too, the revamped rear fenders were different. They were wider and they had higher crowns. However, these fenders were not used on the Plymouth Suburban and station wagon models, which retained their 1949-style rear fenders through the 1952 model year.

Accompanying the re-shaped fenders, when used, were larger taillights. They were positioned lower than in 1949. Their wraparound design gave the cars a smoother profile and also functioned as a safety feature.

In both 1949 and 1950, the Plymouth license plate holder was mounted in the center of the rear deck. However, the mounting was lower on the newer model and Plymouth lettering was added just above the trunk handle.

The chrome trim on the 1950 Plymouth's front and rear fenders appeared to be a bit more streamlined, due to its smoother ends and slightly lower mounting position. The rear fender stone guards were also re-shaped and repositioned.

Interior changes were modest, consisting primarily of altered color toning for the dashboard. There were also minor revisions to the instrument panel control knobs.

In 1949, Plymouth spoke of a "livelier, more powerful, more efficient engine ... the result of a newly designed cylinder head, increased compression ratio (from 6.7:1 to 7.0:1) and a new intake manifold." These changes increased the Plymouth engine's output from 95- to 97-horsepower. The latter rating for 1950 put Plymouth last in what was admittedly a horsepower non-race between Ford with its 100-horsepower V-8 and Chevrolet with its 105-horsepower overhead valve six-cylinder (Powerglide) engine.

The 1950 Plymouths, like their predecessors, were available in two car-lines with significantly different wheelbases. The P-19 models had a 111-inch wheelbase, while the P-20 wheelbase measured 118.5 inches. In contrast, Chevrolet and Ford offered only one chassis size. Nonetheless, the Plymouth's lack of overhang made even the long-wheelbase models shorter than either competitor. The small Plymouth P-19 was 186.5 inches long and the larger Plymouth P-20 was 192.65 inches long. These overall lengths compared to the 115-inch wheelbase Chevrolet's 197.5 inches and the 114-inch wheelbase Ford's 196.8 inches.

In terms of production, Plymouth not only trailed behind Chevrolet and Ford, but faced strong competition from Buick, which had introduced its very popular new Special series in August 1949. While Plymouth still lead in sales charts, Buick actually out-produced Plymouth for the 1950 model year. The calendar-year production figures for the four companies were: 1,520,577 for Chevrolet; 1,187,112 for Ford; 573,116 for Plymouth; and 552,827 for Buick. In contrast, the model-year numbers were: 1,371,535 for Chevrolet; 1,209,549 for Ford; 670,756 for Buick; and 610,936 for Plymouth.

Although not as rare as some other Plymouth models, such as the wooden station wagon, all-steel Suburbans and fastback Concords, the two-door P-20 Special Deluxe Club Coupe is typical of the type of car mid-America bought in the early 1950s. Plymouth ads of the era conveyed the car's conservative image by displaying two- or four-door models filled to capacity with congenial family members. Back then, Plymouth's slogan was "Plymouth builds great cars." Not flashy or sporty, they were simply good, solid, economical transportation like the original Mayflower, which was the company's symbol.

SPECIFICATIONS

Year 1950
Make .... Plymouth
Model .... Special Deluxe
Body style .... two-door Club Coupe
Base price $1,603
Engine .... Inline; L-head; six
Bore x stroke .... 3.25 x 4.375 inches
CID 217.8
Compression ratio .... 7.0:1 (Manual)
Carburetor .... Carter BB 1V
H.P. .... 97 @ 3600 rpm
Wheelbase .... 118.5 inches
Overall length .... 192-5/8 inches
Weight .... 3,041 pounds
Tires .... 6.70 x 15

At the front of the 1950 Plymouth was a redesigned radiator grille with a single center bar. The chrome fender trim appeared to be a bit more streamlined, due to its smoother ends and slightly lower mounting position.

The 1950 Plymouths came in two car-lines with different wheelbases. The P-19 models had a 111-inch wheelbase and the P-20 wheelbase measured 118.5 inches. Illustrated here is a two-door sedan with Deluxe trim.

Although not as rare as some other Plymouth models the two-door P-20 Special Deluxe Club Coupe is typical of the type of car mid-America bought in the early 1950s.

The 1950 Cranbrook fastback was rare. Plymouth ads conveyed the car's conservative image by displaying sedans filled to capacity with congenial family members. Plymouth's slogan was "Plymouth builds great cars."

1956 brought Plymouth's first muscle car

By R. Perry Zavitz

The 1955 introduction of the Chrysler 300, so named because of its horsepower rating, set all other performance cars back on their "wheels." It was a factory hot rod which was a big hit. It not only returned in 1956 with 40 or 55 more horsepower, but also spawned similar type cars in other Chrysler lines.

The DeSoto Adventurer debuted in 1956 with 320 horsepower. Dodge did not have a specific muscle car model, but offered a couple of engines that could be installed in virtually any model. They developed 260 horsepower with a two-barrel carburetor or 295 horsepower with a four-barrel carburetor.

Plymouth introduced a new model called Fury, whose main claim to fame was a 240-horsepower engine. It was the Fury's own engine, not one that was available in any other Plymouth, or even any other Mopar car.

Now, Plymouth was the performance king of the so-called low-priced three. The Fury's engine developed more power and torque than either Ford or Chevrolet offered in any of their sedans until the following year. Chevrolet did not even have an engine as large as the first Fury's until two years later.

The Fury was more than a Belvedere with an extra bunch of horses shoved under the hood. It had a clutch, manual transmission, drive shaft, rear axle assembly, sway bar, springs, shock absorbers, brakes and wheels, all beefed-up to match its motor's brawn.

But, one did not have to get out and get under to tell the Fury from the other Plymouths. The grille was gold anodized. A gold anodized sweep-panel along the sides set it apart from any other car. The Fury name, in unique chrome script on the rear flanks, proclaimed the car's identity to anyone not up on the latest automotive news. All 1956 Furys were two-door hardtops that were finished in eggshell white.

The Fury was luxuriously finished inside with beige and black cloth-and-vinyl upholstery over foam-padded seat cushions and backs. Thick black carpeting underfoot complemented the plush interior.

But, it was performance that really made this Plymouth so unique. It had not been in production long enough to be considered a production car at the time of the 1956 Speed Week trials at Daytona Beach.

However, in NASCAR supervised tests on the beach, the Fury set a new Flying Mile record (for cars with 259 to 305 cubic inches) of 124.01 miles per hour. It set other records, too. One was 82.54 miles per hour for a standing start mile. That was just over two miles per hour better than the previous record, which was set by Cadillac.

The source of all this flurry was the Fury's engine. It was a 303-cubic-inch V-8 made in Windsor. Because of different marketing requirements in Canada, as well as production economics, many Canadian-built Chrysler products were powered by slightly different engines than the same cars in the United States.

The 1956 Dodge Custom Royal is an example. Instead of the 315-cubic-inch 218-horsepower V-8 used in this country, the Canadian version used a 303-cubic-inch 200-horsepower V-8. It was a souped-up version of this Canadian Dodge engine that rocketed Plymouth into prominence with the first Fury.

Precision balanced rods, a high-performance cam, mechanical valve lifters, high-load valve springs, adjustable rocker arms, a high-speed dual-breaker point distributor, reinforced dome-type pistons, a 9.25:1 compression ratio and a larger four-barrel carburetor were some of the different hardware items used in this Dodge 303 engine to produce the Fury's power plant. Output was increased to 240 horsepower as a result.

The combination of this potent engine, the luxuriously finished interior and the practical size of the Plymouth hardtop made a very interesting car. It was a high-performance family car.

The price for the Fury was $2,866. That was a bit steep for a Plymouth 32 years ago, if one looked at only the price tag. But, considering the performance that money bought, it was great value.

SPECIFICATIONS

Year	1956
Make	Plymouth
Model	Fury
Body style	two-door hardtop
Base price	$2,807
Engine	OHV V-8
Bore x stroke	3.81 x 3.31 inches
CID	303
Compression ratio	9.25:1
Carburetor	Carter WCFB 4V
H.P.	240 @ 4800 rpm
Wheelbase	115 inches
Overall length	204.8 inches
Weight	3,650 pounds
Tires	7.10 x 15
OCPG Value	$18,000

The Canadian-made 303-cubic-inch 240-horsepower V-8 powered the 1956 Fury to several new records at Daytona Beach.

The first Fury was a neat combination of performance car and family car.

There were new cast-aluminum road wheels, a performance hood treatments with flat-black finish from windshield to grille and, inside, a center console. A total of 18 exterior colors were available.

1969 Road Runner: no-frills super car

By John A. Gunnell

Plymouth product planner Joe Sturm had a simple way of looking at things. He felt that muscle car buyers fit into five categories and purchased cars accordingly. Sturm knew that price had a lot to do with the buying habits of people in each group.

Professional and hard core amateur racers were willing to spend the most for peak performance. Weekend racers and "executive hot rodders" also had money to spend, but bought looks as much as muscle. The fifth group ... the young street cruisers who hung out at drive-ins ... had lower incomes, but went after the car with the highest performance that their money could buy.

Sturm worked up a chart plotting the quarter-mile elapsed speeds of contemporary muscle cars against their prices. Not unexpectedly, performance went up as the price increased. Several super cars that could do 100 miles per hour in the quarter-mile were available above $3,300. Lower priced cars couldn't match their performance. Sturm decided that Plymouth should make one that did.

To accomplish the goal, he originally planned a car with no frills except a big engine. Chrysler's massive hemi V-8 would be standard, but bright metal trim, floor mats and even a back seat were extras.

As might be expected in Detroit, changes in the plan were made. By the time that the car (the Road Runner) reached Plymouth showrooms, the base engine became a 383-cubic-inch four-barrel V-8 with special modifications to give 335 horsepower. (No other Plymouths got this engine, although Dodge's Coronet Super Bee did).

A back seat *was* included in the production Road Runner, along with black floor carpeting. There were a few other frills like a "beep-beep" horn, too. Still, the Belvedere coupe-based stripper super car did come out with a $2,870 price tag. A hardtop version released at midyear sold for $3,034 however.

The Road Runner's base price included a four-speed gear box, heavy-duty drum brakes and shocks, F70-14 white streak tires, simulated air scoop hood, chrome roof rails, armrests with ashtrays and cartoon emblems showing the road runner cartoon character created by Warner Brothers Films.

As for performance, stock Road Runners came in just shy of the 100 miles per hour goal. The top quarter-mile terminal speed achieved by Plymouth with the base engine was 98 miles per hour. However, production of the car still hit 45,000, way above the 2,500-unit projection.

For 1969, the Road Runner returned with minor styling and trim alterations and gained a convertible model. A change was made under the hood where the beep-beep horn was painted light purple and dressed up with a decal. The reason for this was that many muscle car freaks had asked to see the unusual horns when the 1968 models came out. However, they found the plain black devices uninspiring.

Base Road Runner power plant for 1969 continued to be the 383-cubic-inch, 335-horsepower V-8, but several new options were made available. In addition to a four-barrel Super Commando V-8 with 440 cubic inches and 375 horsepower offered in 1968, a new mid-1969 choice was the "440 Six-Pak" engine with 390 horsepower. The 426-cubic-inch, 425-horsepower hemi remained a $700-plus option and a slew of new rear axles (high-performance, super-high-performance, Trak-Pak and Super-Trak-Pak) could be had with specific engine/transmission combinations.

For dress-up, there were new cast-aluminum five-slot road wheels and performance hood treatments (with flat-black finish from windshield to grille). Inside, buyers could now order a center console, plus a variety of power equipment options. Red streak tires became standard. Ten new exterior paint colors were offered, making a total of 18 (five only on Six-Pak engined cars).

According to published road tests, the 440 Six-Pak cars were the fastest of the 1969 Road Runners. *Super Stock* magazine put one of these through the quarter-mile in 13.50 seconds at 109.31 miles per hour. Chrysler's Dick Maxwell drove a hemi Road Runner the same distance in 13.59 seconds at 105.63 miles per hour.

SPECIFICATIONS

Year 1969
Make .... Plymouth
Model .... Road Runner
Body style .... two-door hardtop
Base price $3,083
Base engine .... OHV V-8
Bore x stroke .... 4.25 x 3.375 inches
CID 383
Compression ratio 10.0:1
Carburetor .... Carter AVS 4V
H.P. .... 335 @ 5200 rpm
Wheelbase .... 116 inches
Overall length .... 202.7 inches
Weight .... 3,450 pounds
Tires .... 8.25 x 15
OCPG Value $17,000

The 1969 Road Runner. There's still only one place to catch it.

The 1969 Road Runner.
If you want a high-performance car, Road Runner is one car to think about.

The Road Runner is a real performer. But not because it costs a lot of money. It doesn't.

It comes in three models, with a standard 383 cubic inch V-8. A 4-barrel carburetor. An unsilenced air cleaner. And dual exhaust trumpets.

A 4-speed transmission with Hurst Linkage. A high lift cam. And popular Red Streak Wide Boots.

Options include a tachometer, and our new 160-position driver's adjustable bucket seat that does everything a power seat

does. At roughly half the cost. Another new option: functional hood scoops, or "Air Grabbers."

Now there is a larger, full-color Bird on the deck lid, doors and instrument panel. Plus a new deluxe steering wheel—with the famous Bird perched right on the hub.

This year's Road Runner comes in eighteen exterior colors. With broad black sport stripes on the hood, optional.

Pity the poor coyote.

If Road Runner doesn't baffle him with numbers, he surely will with plumage. "Beep-Beep!"

You can only catch the Road Runner at your Plymouth Dealer's. That's the place, and 1969's the time to . . .

Look what Plymouth's up to now.

Watch AFL Football and The Bob Hope Special on NBC-TV

"You can only catch the Road Runner at your Plymouth dealer," the ads said. Base 1969 power plant was the 383-cubic-inch V-8 with a four-barrel carburetor and 335-horsepower. Several new power train options were offered.

231

For 1969, the Plymouth Road Runner coupe returned with minor styling and trim alterations. There was also a new convertible model. Red Streak Wide Boot tires were standard equipment.

A cartoon character badge was indicated that the car could fly down the road. Super Stock magazine found the 440 Six-Pak cars fastest of all 1969 Road Runners in the quarter-mile at 13.50 seconds and 109.31 miles per hour.

If bigger and better reflected the shedding of a depression mentality, Pontiac went in that direction in 1937. This Deluxe Six convertible coupe was considerably larger than its 1936 counterpart.

1937 Pontiac was a great car for a great year

By John A. Gunnell

If 1936 was a good year for Pontiac with 176,270 total vehicle assemblies, 1937 was a great one. The Pontiac, Michigan company wound up adding 286,189 units to General Motors' production total by the end of that model year. The figures included 179,244 six-cylinder models and 56,945 straight eights.

The car-lines were cut from three to two. The Master Six series was no longer available. Remaining were the Deluxe Six and Deluxe Eight series. Cars in both lines had wheelbases five inches longer than the previous season: 117 inches for sixes and 122 inches for eights.

No longer used by Pontiac was General Motors' small "A" body. Some said that the bigger size of the 1937 cars made it clear that the depression was ending. Buyers had developed a positive attitude towards the future and wanted bigger and better things.

Better things for 1937 included all-steel body construction, Hotchkiss drive (allowing a lower riding height) and larger, more powerful engines. The inline six-cylinder engine went up to 85 horsepower, while the inline eight-cylinder power plant was advertised at 100 horsepower.

While not a high-performance car in those days, the Pontiac was capable of 86 miles per hour speeds and could go from a standing start to 60 miles per hour in about 19 seconds. Those figures were recorded during an extensive test conducted by the British magazine *Autocar* at the Brooklands race course.

If bigger and better reflected the shedding of a depression mentality, Pontiac went even a little further in that direction by the time that April 1937 rolled around. That was the month in which the company introduced a new midyear model. It was a four-door convertible sedan.

Pontiac used the same bodies for both its six- and eight-cylinder cars. However, the front end sheet metal varied according to engine size and wheelbase. The Deluxe Eights, with their longer stance, had longer front end parts.

Using the body in common made it possible to offer Convertible Sedans in both lines. In fact, the Deluxe Eight was priced at only $50 more than the Deluxe Six.

The Convertible Sedan was a great new product for what turned out to be a great Pontiac year. Unfortunately, the new model did not do that great in the marketplace. According to records, only 1,266 cars of this body style were produced by Pontiac during the 1937 model year. The few surviving today are cherished by collectors.

SPECIFICATIONS

Year	1937
Make	Pontiac
Model	Deluxe Six
Body style	Convertible Sedan
Base price	$1,197
Engine	L-head; inline; 6
Bore x stroke	3-7/16 x 4 inches
CID	222.7
Compression ratio	6.2:1
Carburetor	1V
H.P.	85 @ 3520 rpm
Wheelbase	117 inches
Overall length	193.06 inches
Weight	3,375 pounds
Tires	16 x 6.00
OCPG Value	$26,000

SPECIFICATIONS

Year	1937
Make	Pontiac
Model	Deluxe Eight
Body style	Convertible Sedan
Base price	$1,235
Engine	L-head; inline; 8
Bore x stroke	3-1/4 x 3-3/4 inches
CID	248.9
Compression ratio	6.2:1
Carburetor	2V
H.P.	100 @ 3800 rpm
Wheelbase	122 inches
Overall length	198.06 inches
Weight	3,505 pounds
Tires	16 x 6.50
OCPG Value	$28,000

The Convertible Sedan was a great new product in what turned out to be a great Pontiac year. Unfortunately, this model did not do that great in the market. The few surviving are cherished by collectors. (Pontiac photo)

Pontiac's 1937 models had a raised hood line. Silver Streak moldings ran down the hood in "waterfall" fashion. A narrower radiator grille was another styling feature that showed up on the new Convertible Sedan. (Pontiac photo)

Using the body in common made it possible to offer Convertible Sedans in both of Pontiac's 1937 car-lines. According to records, only 1,266 Convertible Sedans were produced by Pontiac during the 1937 model year.

Big time 1941 Pontiac coupe

By John A. Gunnell

Pontiac hit the big time in 1941. A new line of cars ... dubbed the "Torpedo Fleet" ... made the company America's largest producer of medium-priced cars for the first time. New styling, new features and new models pushed sales for the year over 330,000 vehicles.

The increase of 113,000 deliveries over 1940 went mainly to bread-and-butter models in the lower-priced car-lines. However, sales were very likely helped along by a marketing program that gave the least expensive 1941 more of a big-car image.

On September 15, 1939 Pontiac introduced two top-of-the-line Torpedo fastbacks. They shared the General Motors corporate C-body with large Buicks, Oldsmobiles and Cadillacs. Although less than 32,000 were built that season, the first Torpedos created the impression that Pontiac was marketing luxury cars for just above "everyday" prices.

To reinforce the impression of great value for the money, Pontiac renamed all of its 1941 models Torpedos. This helped boost sales, since buyers believed they were getting more car for their money. In reality, the higher-priced cars (now called Custom Torpedos) saw a sales decline of nearly 6,500 units. This drop came despite the addition of new models in the series. One, the Custom Torpedo Sport Coupe, is quite rare today.

Unlike Buick's fastback-only line, the new C-bodied coupe was a conventional notchback type that looked very similar to other five-window Pontiac coupes. However, its trunk was proportioned differently and it had individual moldings around each side window.

A comparison between the dimensions of the Custom Torpedo and the other Pontiac sport coupes is interesting. It proves that the high-priced model did not utilize space very efficiently. It had the same 122-inch stance as the B-bodied Streamliner Torpedo Sport Coupe, although both of these had a three inch longer wheelbase than A-bodied Deluxe Torpedos. Length of the big coupe was four inches more than a similar Streamliner and 10 inches more than a similar Deluxe Torpedo. However, the B-bodied Streamliners were actually widest and the A-bodied Deluxes were tallest.

Interior measurements tell the same story. In most regards, the B-bodied Sport Coupe actually had the most passenger room. Only in width and depth of front and rear seats was the C-bodied version bigger and, then, just by fractions of an inch.

Size-wise, the large Pontiac's most noticeable difference was in the depth of its trunk. It was eight inches longer than the B-body's trunk and 16 inches longer than the C-body's trunk. That is what made the Custom Torpedo Sport Coupe look really different.

Pontiac selling features for 1941 included new brakes; new clutch controls; new adjustable sun visors; standard dual rear lamps with automatic stop signals; rubber-insulated step-type body mountings; improved rust-proofing; new bridge-type frames; improved Multi-seal hydraulic brakes; additional axle ratio options (from 3.9:1 to 4.55:1); and a new built-in full-flow permanent oil cleaner in both six- and eight-cylinder engines.

To go with the luxury image, Custom Torpedo Sport Coupes had rich interior furnishings and seats covered in a thick worsted wool cloth in a two-tone blend with twin pinstripe.

Solid paint colors available in 1941 included Marlboro Blue; Parma Wine; Streamline Gray; Tropic Blue; Allandale Green; Indiana Beige; and Taffy Tan. Six two-tone combinations were also available: El Paso Beige over Indiana Beige; Paddock Gray over Marlboro Blue; Silver French Gray over Streamline Gary; Thetis Green over Allandale Green; French Copper over Taffy Tan; and Santone over Parma Wine.

SPECIFICATIONS

Year	1941
Make	Pontiac
Model	Custom Torpedo Eight
Body style	two-door Sport Coupe
Base price	$1,020
Engine	L-head; inline eight
Bore x stroke	3-9/16 x 4 inches
CID	248.9
Compression ratio	6.5:1
Carburetor	Carter 2V
H.P.	103 @ 3500 rpm
Wheelbase	122 inches
Overall length	211.5 inches
Weight	3,325 pounds
Tires	6.50 x 16
OCPG Value	$13,000

The 1941 Torpedos created the impression that Pontiac was marketing luxury cars for just above "everyday" prices. However, the big C-bodied Custom Torpedo Sport Coupe was a bit pricier than your standard Pontiac.

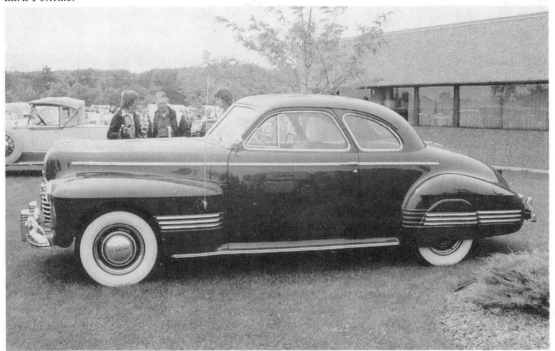

Pontiac's 1941 Custom Torpedos actually saw a sales decline of nearly 6,500 units and the Sport Coupe in the series is quite rare today. Those rear fender skirts, complete with triple chrome bars, were optional.

Comparing the Custom Torpedo with other coupes shows its inefficient space utilization. Its trunk is eight inches longer than the B-body trunk and 16 inches longer than the A-body trunk and makes it look totally different.

Adventurer Lowell Thomas preferred the 1941 Pontiac Torpedo four-door sedan. Solid colors for 1941 were Marlboro Blue; Parma Wine; Streamline Gray; Tropic Blue; Allandale Green; Indiana Beige; and Taffy Tan. (Pontiac photo)

1963 Grand Prix: plain was beautiful

By John A. Gunnell

"Here's what the other '63s wished they looked like," boasted Pontiac's ads. It was more than just hype. After a spectacular 1962, Pontiac featured all-new sheet metal in what its designers called the "venturi" theme. Most people referred to it as the Coke bottle shape.

Perhaps its best application was to the 1963 Grand Prix. Instantly popular upon its introduction the year before, this second-year version, with its newly stacked headlights and no side trim, was a spectacular success. Almost 73,000 of the two-door hardtops left the factory, a whopping 140 percent increase from 1962.

It was quite a departure for a top-of-the-line model to be almost devoid of chrome. This was once the sign of a cheaper car. But, the 1963 Grand Prix looked anything but cheap. It exuded good taste from every angle. Along with the unusual vertical headlights, a concave rear window and partially hidden taillights were styling innovations that worked perfectly on this most expensive two-door hardtop ($3,490) in the Pontiac line-up.

For that price, the Grand Prix buyer got the considerable standard equipment found on Bonnevilles, plus special Morrokide upholstery, wood-grained steering wheel and dash trim, front bucket seats and center console with vacuum gage. Oddly enough, automatic transmission was not included. Three-speed manual units were standard on *all* Pontiacs in 1963 and 5,157 Grand Prixs were so-equipped.

The standard power plant was Pontiac's 389-cubic-inch V-8, which developed 235 horsepower when connected to the synchromesh manual transmission, but a more impressive 303 horsepower when teamed with Hydra-Matic. The tri-power induction option on the 389 boosted its horsepower to 313. The Grand Prix could also be ordered with two versions of the 421 H.O. V-8, developing 353 or 370 horsepower.

SPECIFICATIONS

Year	1963
Make	Pontiac
Model	Grand Prix
Body style	two-door hardtop
Base price	$3,490
Engine	OHV V-8
Bore x stroke	4.06 x 3.75 inches
CID	389
Compression ratio	8.6:1 (Manual)
Carburetor	Carter AFB 4V
H.P.	235 @ 3600 rpm
Wheelbase	120 inches
Overall length	211.9 inches
Weight	3,915 pounds
Tires	8.00 x 14
OCPG Value	$10,000

"If you think the Grand Prix is just a big, beautiful, elegant car, you obviously haven't driven it," warned this ad, an allusion to the car's high-performance charateristics.

If you drive this car, you're going to have to get used to sharing its beauty with a lot of spectators. You really can't blame people for clustering around this new Grand Prix, can you? And it shouldn't bother you much, anyhow. You'll be too busy enjoying its utterly civilized road manners. A wider Wide-Track does it, along with a calm, collected ride. The vigorous Trophy V-8 humming away under the hood helps, too. But instead of just sitting there reading a list of GP joys, why don't you sample the whole package firsthand? Your Pontiac dealer's the man with the keys. He'll let the GP do its own selling—but we warn you: it's the most effective sales talk you ever heard. Make plans to listen in soon.

PONTIAC MOTOR DIVISION · GENERAL MOTORS CORPORATION

GP

PONTIAC GRAND PRIX

"If you drive this car, you're going to have to get used to sharing its beauty with a lot of spectators," Pontiac teased in another ad from the series rendered to show the sporty Pontiac in European settings.

"It may be some time before you see this car without a lot of people crowded around it, so be our guest,"
Pontiac invited in this 1963 Grand Prix advertisement.

If you don't want to be looked at wherever you go, don't go in this!

We're nothing if not anxious to help you lose your heart to our new Grand Prix. In fact, we'll not only let you *look* to your heart's content, we'll tell you a bit about it. You'll find a pair of deep bucket seats up front, separated by a handy control console. A throttle that connects you to our 389-cubic-inch, 303-hp Trophy V-8 (standard, by the way). Rich carpeting that may never wear out, as far as our testers can tell right now. Self-adjusting brakes (like all Pontiacs). But we'd better save some of the GP's wonders for our dealers—we figure you ought to be on your way to one of them just about now.

PONTIAC GRAND PRIX

The arty ads were classy and less expensive than shipping a Grand Prix around the globe. "If you don't want to be looked at wherever you go, don't go in this," was the theme here. Standard engine was Pontiac's 389-cubic-inch V-8.

The distinctive Trans Am hood incorporated dual functional air scoops on either side of a central crease. Blue vinyl stripes streaked back from each air scoop. (PMD)

Ordering a 1969-1/2
Firebird Trans Am

By John A. Gunnell

One of the hottest items in the collectible pony car market is the first Trans Am, which appeared as an option-created car in mid-1969. This model-option came in coupe and convertible styles. Today, in top condition, they can cost as much as $14,000 and $22,000, respectively.

Originally conceived of as a sedan-racer, the Trans Am was to have a special, low-compression, small cubic-inch, high-horsepower engine that qualified for Sports Car Club of America (SCCA) Trans-Am Series racing. However, the equipment and handling modifications were readied before the engine was completed, so the car was released with other power plant options. These were also true high-performance engines, although too large in cubic inches to meet SCCA racing guidelines.

Ordering a 1969-1/2 Trans Am started with a sales contract for a base Firebird. Standard equipment on this car (either coupe or convertible) included all safety features regularly found in General Motors cars, a 175-horsepower overhead cam six, E70-14 black fiberglass Wide-Oval tires, front bucket seats, camera-case grained dashboard, carpeting, left-hand mirror, side marker lights and front seat head restraints.

Dealer cost on the coupe was $2,161.89 and federal excise tax was $159 making the total $2,320.89. It sold for $2,853 at retail, including a $25 dealer preparation charge. The convertible cost $2,338.47 plus $171 federal excise tax, for a total of $2,509.47. It retailed at $3,083.00. None of these figures included accessories.

The Code 322/UPC WS4 Trans Am "performance and appearance package" was an option. Basics listed for the package were a 400-cubic-inch four-barrel Ram Air V-8; power steering, front disc brakes, Ram Air inlet hood, Ride and Handling springs and shocks, F70-14 whitewall fiberglass tires, heavy-duty stabilizer shafts and bushings and functional front fender engine compartment air extractors. The package came with four different installation prices, but in round figures it cost $1,100, over base price, to make either one of the base Firebirds into a very "plain-Jane" Trans Am.

The first T/A was just under 16 feet long and was pure Pontiac from stem-to-stern. Like all 1969 Firebirds, it had a new nose treatment (with a boxy split grille); exposed, square-bezel headlights; Endura headlamp surrounds; and crease-line front fender sculpturing. The overall design was highlighted by a Mustang-inspired long hood/short deck theme and Coke-bottle rear fender kick up. Coupes had true notchback roof styling. Buried in the flat rear panel were tapering and stacked slits housing the taillamps, with two slits on either side.

The distinctive Trans Am hood incorporated dual functional air scoops on either side of a central crease. Blue vinyl stripes streaked back from the inlet openings to the cowl, matched by similar markings on the roof and air-foiled rear deck lid. A variation seen here was stripes running under the spoiler on some cars, across it on others. Decals carried the Trans Am name in blue block letters at the lower front fender tips and across the full-width air foil, or spoiler.

The Trans Am interiors were not much different than those provided in other Firebirds. Cars ordered with optional Firebird Custom interior trims got wood-look dash inserts (walnut or Carpathian Elm burl patterns) and a handful were upholstered in gold leather and expanded Morrokide trim combinations.

If one thing stood out in overall design, it was the fact that 1969 Trans Am blended together a number of different, not normally compatible, design elements and did it well. For example, it had sedan/racer characteristics, California customizing tricks and many street rodding touches. Individually, the pieces did not seem to go together, but somehow they fit just perfectly.

The Trans Am was sometimes criticized for its road handling manners, but never on its straight-line acceleration abilities. Quarter-mile times registered by magazines of the day ranged from 14 to 14-1/2 seconds and terminal speeds were at the century mark.

SPECIFICATIONS

Year	1969-1/2
Make	Pontiac
Model	Firebird Trans Am
Body style	two-door hardtop
Base price	$3,556
Engine	OHV V-8
Bore x stroke	4.12 x 3.75 inches
CID	400
Compression ratio	10.75:1
Carburetor	Rochester 4V
H.P.	335 @ 5000 rpm
Wheelbase	108.1 inches
Overall length	191.1 inches
Weight	3,080 pounds
Tires	F70-14 RWL
OCPG Value	$14,000

The Trans Am had a new nose treatment with a boxy split grille; exposed, square-bezel headlights; Endura headlamp surrounds; and crease-line front fender sculpturing. (PMD)

Quarter-mile times registered by magazines of the day driving the first Trans Am ranged from 14 to 14-1/2 seconds and terminal speeds were at the century mark. (PMD)

Decals carried the Trans Am name in blue block letters at the lower front fender tips and across the full-width air foil, or spoiler.

1962: When the Cobra struck

By R. Perry Zavitz

Though tall and lanky and born in Texas (in 1923), Carroll Shelby was not a typical Texan. He was not a wealthy oil man, yet he did work briefly in the oil fields for his father-in-law. He was not a cowboy or rich cattle baron. He did begin raising chickens, but disease wiped out his flock and left him broke. At one time he ran a small fleet of ready-mix concrete trucks, but his main interest was in auto racing.

Beginning with a borrowed MG, he began participating in local races and did it with rather consistent success. Racing would take him further afield to other states, such as Florida for Sebring and Utah for Bonneville. Many were events sanctioned by the Sports Car Club of America. His success also led him to other countries:. Mexico for the Carrera Pan Americana, Argentina, England, Italy and France, to name some. At LeMans, in 1959, Carroll Shelby drove an Aston-Martin to victory, ousting Ferrari from the checkered flag position.

Health began interfering with his racing career. He had heart problems. Racing in the top international events was not on his doctor's list of recommended therapeutic activities.

An alternative for Shelby seemed to be the development of a high-performance car. That was exactly what he began concentrating on. While in Europe and elsewhere, he became quite familiar with the various cars and their respective strengths and weaknesses. Cad-Allards and Nash-Healeys were impressive, but he believed they could be improved upon. Coincidently, at this time, AC was facing a crisis when their engine supplier, Bristol, announced it would discontinue making auto engines. AC was prepared to use Ford of Britain's six-cylinder motor as a successor.

Another coincidence was Shelby's discovery that the Ford Motor Company, here in the states, was well ahead in developing a new light weight V-8 engine. He contacted AC, suggesting the use of this new Ford V-8 power in their AC Ace model. AC responded with interest in the matter. Meanwhile Shelby, through contacts he recently happened to make at Ford, obtained a pair of the new 221 cubic inch thin-wall V-8 motors.

Ford had been working on this new cast-iron engine block since 1958, believing that this was the right road to lighter motors, rather than aluminum. History has proven Ford to be right in this regard. Remember the flurry of aluminum engines in the 1960s? Where are they now?

Ford's new engine tipped the scales at close to 450 pounds. It was only 15 pounds heavier than the Bristol engine AC had been installing in its sports car. Carroll Shelby found that this Ford V-8, measuring 21 inches long, 16 inches wide and just nine inches high, slipped quite easily into the AC chassis.

When Shelby got delivery of his first engine-less AC body, he and his men installed the Ford engine. They worked in the small speed shop of his friend Dean Moon in Sante Fe Springs, California. It is reported that Shelby was road testing the car just eight hours after the AC arrived.

For a moment, let's take a quick glance at AC's background. AC was a car maker in Surrey, England, whose history can be traced back at least to 1908. The initials stand for Autocarriers, which was its name until 1922. In 1953, AC introduced a compact roadster called the Ace. Its aluminum body had just the right blend of curves to be attractive. More curves would have made it look fat and bulbous. Fewer curves would have made it flat and uninteresting.

The Ace used a motor supplied by Bristol, which was a British airplane maker that got into car production after World War II. Neither the Ace's body or engine had changed much during the nine years since its introduction. The engine, originally of BMW design, was a two liter six-cylinder mill rated at 100 horsepower.

Shelby's new hybrid had a 4.3 liter V-8 that developed 260 horsepower. The engine he used was not the 221 cubic inch Ford motor, but one of two variations of it, which was just now coming on stream. The over-bored edition of the 221 had 260 cubic inches. This was used by Ford for a V-8 Falcon called the Sprint. Although it developed 164 horse-power, there was a 260 horsepower Monte Carlo option.

The short 2-7/8 inch stroke remained. It was one of the shortest strokes ever used in a production American car. The contemporary Corvair and Falcon had shorter stroke six-cylinder engines. Among V-8s, only the Pontiac Tempest, Olds F-85 and Buick Skylark had slightly shorter strokes. Regardless of the significance of that, however, it took little persuasion to get Carroll Shelby to insert this potent 9.2:1 compression, four-barrel carburetor engine in his car.

In order to withstand the strain of all that extra power, Shelby made some modifications to the AC. For example, a Ford four-speed manual transmission was appropriately linked to the engine. Other changes were made to the English chassis to overcome foreseeable problems or to quickly correct unforeseen problems that arose. The AC's tubular space frame used some heftier pipes.

Shelby would have preferred to use a fiberglass body, but could not wait for one to be developed. So the Ace's clean-looking aluminum body was used with a couple of modifications. The fenders were given slightly protruding lips to accommodate the larger 6.40 x 15 (front) and 6.70 x 15 (rear) Goodyear tires. (Carroll Shelby was a Goodyear tire distributor.) These fender lips had flat edges. They were much better looking than the flared out fenders on some later models.

The other body change was the replacement of the AC emblem above the grille with a Shelby Cobra emblem. Evidently, calling his car the Cobra was an idea long in the mind of Carroll. Venomous snakes are often on the minds of some Texans, but the cobra is not native to North America. The name had been used by the little Crosley, but not for its cars. It was an acronym for the COpper BRAzed method of fabricating its engine from sheet steel. But, that type of engine disappeared by 1950 and the Crosley itself had disappeared in 1952.

There are some discrepancies over exactly what the Shelby-built Cobra should be called. The body supplier called it the AC Cobra. The engine supplier called it Ford Cobra, but did settle for just Cobra because of the "powered by Ford" explanation appearing on the Cobra nameplate. The producer preferred the Shelby-Cobra title. Those wanting to be diplomatic and fair called it the AC Ford Cobra by Shelby. Take your choice.

After all the modifications were made and the Cobra was ready for the road, it had a curb weight of 2,385 pounds. With the robust 260-horsepower power plant, Shelby's car was an astounding performer. The earliest published test appeared in *Sports Car Graphic*. They reported a 0 to 60 acceleration time of a mere 4.1 seconds, a standing start quarter-mile in 12.9 seconds at 114 miles per hour and a top speed two-way average of 152 miles per hour.

Some think those figures are a bit suspect. If a car can do 0 to 60 in just 4.1 seconds, then it should do a quarter-mile at a faster speed than 114 miles per hour. And the top speed of over 150 just seemed unbelievable. Subsequent test report's showed similar results, which the skeptics say were just other magazines avoiding the embarrassment of reporting less impressive performance. Be that as it may, the Cobra's performance was indeed awesome. And it was enough to give the Corvette fits.

Comparing the Cobra and the Corvette is like comparing apples and oranges, some critics claim. Agreed. The two cars are much different. Yet, apples and oranges are compared.

The best performance figures we have been able to find on the 1962 Corvette were published by *Motor Trend* using a 360-horsepower fuel-injected model with four-speed stick shift. They claimed 0-to-60 in 5.9 seconds, a standing start quarter-mile in 14.9 seconds with a trap speed of 102.5 miles per hour and a 132 miles per hour top speed. That all falls quite short of the Cobra's reported performance. Of course, the Corvette was a heavier car with a curb weight of 3,137 pounds.

One advantage the Corvette had was its price. With a 360-horsepower fuel-injected engine and four-speed manual transmission, the Corvette had a $3,887 price tag. The base Cobra cost a hefty $5,995.

There were only about 75 of the 260-cubic-inch Cobras produced, when Shelby began using Ford's larger 289-cubic-inch engine. Those used in the Cobras were rated at 271 horsepower. The majority of Cobras had the 289 engine. Even many of the 260 engines were replaced with 289s by owners.

The ultimate hybrid came in 1964, when Shelby began producing the Cobra 427. It was the result of stuffing Ford's 427-cubic-inch engine in the AC chassis. Ironically, the 289 and 429 powered Cobras did not have quite as good performance as the original 260 models (if the 260 test results are correct). For instance, 0 to 60 in 5.6 seconds for the 289 motor recorded by *MoTor* and 4.8 seconds for the 427 registered by *Autocar*.

From its humble beginnings 25 years ago, Shelby's Cobra became an overnight success. That success has not faded noticeably over the past quarter century. Has there ever been a postwar North American car that has triggered so many replicas as the Cobra?

SPECIFICATIONS

Year	1962
Make	Shelby
Model	Cobra
Body style	two-door roadster
Base price	$5,995
Engine	OHV V-8
Bore x stroke	3.8 x 2.87 inches
CID	260
Compression ratio	9.21:1
Carburetor	Holley 4V
H.P.	260 @ 6500 rpm
Wheelbase	90 inches
Overall length	151.5 inches
Weight	2,385 pounds
Tires	6.40 x 15
SCAC Value	$190,000

Carroll Shelby's "Cobra Caravan" promoted his company's full product line in 1965 which included (left to right) the 427 Cobra roadster, the Ford GT-40, the Daytona Super Coupe and the Shelby-Mustang. (Courtesy Carroll Shelby)

Those wanting to be diplomatic and fair called it the AC Ford Cobra by Shelby. Whatever you called it, the high-performance roadster was in a class of its own. (Courtesy Carroll Shelby)

Styling steadied Studebaker

By John A. Gunnell

In March 1933, Studebaker slid into receivership. Then, company president Albert Erskine committed suicide. Harold Vance and Paul Hoffman faced the job of pulling the South Bend, Indiana firm up by its boot straps.

Studebaker stock in Pierce-Arrow was sold off for $1 million. Then, $100,000 was spent to launch a "Studebaker Carries On" campaign. Streamlining, in terms of the product mix, as well as in terms of styling, helped to cut operating costs and drive up sales.

Studebaker's styling for 1934 was sleeker and much more modern. The company's "tommorrow's-car-today" appearance included a slanted V-type radiator, aerodynamic body feature lines, deeply skirted fenders and rakish belt line moldings.

Changes for 1935 further emphasized futuristic styling. It was headlined by a longer, much narrower grille, multiple horizontal hood louvers, and more bullet-shaped headlamp buckets.

Three series called Dictator, Commander and President were offered. Wheelbases varied according to series: 114 inches; 120 inches; or 124 inches in the same respective order. A 205-cubic-inch, 88-horsepower L-head six-cylinder engine powered Dictators. Commanders used a 250-cubic-inch, 107-horsepower straight eight. Presidents shared the same basic eight, but with horsepower rating raised to 110 horsepower.

All 1935 Studebakers featured a three-speed manual transmission, single-plate dry-disc clutch, shaft drive and semi-floating rear axle. Freewheeling was optional. On Presidents, automatic overdrive transmission was standard. Four-wheel hydraulic brakes were used on all models that season. Tires came in a variety of sizes: 16 x 6.00 or 17 x 5.50 for Dictators; 16 x 6.50 for Commanders; and 16 x 7.00 for Presidents. They were all mounted on steel artillery spoke wheels.

These cars were introduced on December 28, 1934. Innovations included new planar front wheel suspension (except on base Dictators), automatic overdrive, hydraulic brakes on Dictators and provisions for mounting radios in all cars. A soon-to-be-familiar name, Champion, was used in some 1935 Studebaker advertisements.

Studebaker offered 14 versions of the 1935 Dictator. Each of them came as a 1A (with solid front axle) or a 2A (with planar front suspension) model. Prices for 1A Dictators ranged from $695 for a Business Coupe to $895 for a Regal Land Cruiser sedan. Planar front suspension was $25-$35 extra. Regal and Custom trim levels were included in the model line-up.

In the Commander series, prices ranged from $925 for the coupe to $1,130 for the high-trim Regal Land Cruiser. There were 12 models and all had planar front suspension.

Also offering 12 models was the President series. This top-of-the-line range began with the $1,245 coupe and peaked with two $1,445 cars called the Regal Land Cruiser and the Regal five-passenger Berline.

To promote sales in 1935, Studebaker did some wild stunts. They included one in which a sedan was driven over the railroad ties of the world's largest electric railroad bridge in Berrien Springs, Michigan at 60 miles per hour.

Styling changes combined with management improvements put Studebaker back on a steady course. The media hype seemed to work, too. By March 1935, the company emerged from receivership. With Paul Hoffman as president and Howard Vance as chairman of the board, Studebaker placed 11th in the ranks of automakers for the model year with sales of approximately 36,000 units.

SPECIFICATIONS

Year	1935
Make	Studebaker
Model	President
Body style	four-door Regal Land Cruiser Sedan
Base price	$1,430
Engine	L-head; inline; eight
Bore x stroke	3-1/16 x 4-1/2 inches
CID	250
Compression ratio	6.5:1 (Manual)
Carburetor	Stromberg EE-1
H.P.	110 @ 3600 rpm
Wheelbase	124 inches
Overall length	211.9 inches
Weight	3,900 pounds
Tires	7.00 x 16
OCPG Value	$11,300

Two warmly dressed women eyeball the 1935 President Custom sedan. Changes for 1935 emphasized futuristic styling with a longer, much narrower grille, multiple horizontal hood louvers, and more bullet-shaped headlamp buckets.

Studebaker Presidents 8s shared the same basic engine as 1935 Commanders, but power was raised slightly to 110-horsepower. On Presidents automatic overdrive transmission, four-wheel hydraulic brakes, and 16 x 7.00 tires were standard.

Also available with similar overall styling but plainer trim and fewer features was the Dictator. This fellow is showing his new 1935 Dictator five-passenger sedan to a pair of female admirers.

Studebaker was first in '55

By John A. Gunnell

Studebaker was the first American automaker to introduce its 1955 models. As a result of their early showroom arrival, *Car Life* magazine grabbed a President State hardtop for its initial "Behind the Wheel" road test that year.

Staff writer G.M. Lightowler described his lemon yellow President as a "sporty sedan (sic) with high performance." Appearance-wise, he liked the car, but criticized its "butter knife" side trim and gaudy fender top moldings suggesting that automakers might have to go into the chrome cleaner business if they kept adding "tinsel" to their cars.

The basic design drew high praise. Photographically, the magazine compared it to an Austin-Healey sports car to emphasize its low hood height. "We would still like to see such a trim basic body in a convertible style," the caption stated. (Probably, a lot of modern collectors would have liked that, too.)

The President's sleek body drew many admiring looks at traffic lights. However, it proved somewhat drafty when driven in bad weather and rain water also leaked into the car near the drafty spots. Lightowler did indicate that the leaks could probably be repaired fairly easily.

Interior trim came in for lots of positive comments, as did visibility, driver comfort and general fit and finish. The seats were upholstered in green vinyl, the floors had light green woven nylon coverings, and the headliner was white plastic material. The high-level President State model tested had a new rear seat center armrest. Its new front bench seat was higher and wider and front bucket seats were optional, which was fairly unusual for a domestic car of 1955 vintage.

Least liked inside the car was the instrument panel, which had some distracting design elements and cheap-looking gold finish. However, most of the controls, knobs and switches were found to be well laid out. Key-start ignition replaced a starter button under the clutch.

Driving the car was called "exhilarating." It had the hot "Wildcat" V-8 and automatic transmission. Low gear acceleration (up to as high as 56 miles per hour) was impressive. Top speed was more than 100 miles per hour. Road handling, cornering characteristics and the suspension all got very high marks. So did the new Firestone tubeless tires.

Stopping and steering came in for criticism. The problem with the brakes was actually one of pedal position. It was mounted in a "definitely bad" location, too high for convenient operation. The power-assist on the steering caused the car to oversteer. Lightowler suggested that it wasn't necessary or advisable to order power steering for a high-performance car like the Studebaker President. He did, however, miss other extras, saying that a car with a $2,600 price tag should have some as standard features. Surprisingly absent were turn signals, back-up lights, a cigarette lighter and an outside rearview mirror.

Studebaker's retention of six-volt electrics also lost it a few "brownie points," although no trouble was experienced in starting the test car one frosty morning. It did, however, get overheated while waiting in a traffic jam caused by a crowd going to a football game.

In more modern times, as a collectible special-interest car, the 1955 Studebaker President State two-door hardtop makes a nice everyday driver. It's plusher and sportier than the standard President sedan, but not quite as hard-to-find as the usually trailer-to-shows President State Speedster hardtop.

Surprisingly, the Speedster (with production of 2,215 units) is not that much rarer than the State hardtop with its 3,468 assemblies. In fact, the Speedster is the same car with minor trim upgrades and nearly all the factory options added. Despite its two-tone finish and special engine-turned dashboard, the Speedster had the same V-8 as the President State and more of a true sports car feeling. *Car Life*'s test driver probably would have preferred the plainer President, after getting a look at the yellow-and-green Speedster with even more unnecessary brightwork.

SPECIFICATIONS

Year	1955
Make	Studebaker
Model	President State
Body style	two-door hardtop
Base price	$2,456
Engine	OHV V-8
Bore x stroke	3-9/16 x 3-1/4
CID	259.2
Compression ratio	7.5:1
Carburetor	Carter WCFB 4V
H.P.	175 (early) or 185 (late) @ 4500 rpm
Wheelbase	120.5 inches
Overall length	204.4 inches
Weight	3,175 pounds
Tires	7.10 x 15
OCPG Value	$14,000

The 1955 President State hardtop was low-slung and sporty. Car Life magazine compared it to an Austin-Healey 3000 sports car to emphasize its extremely low hood height. Two-tone finish was popular. (Old Cars photo)

The President State was available in solid colors, too. A massive chrome grille was new. Its sleek body drew many admiring looks at traffic lights, but had some drafty window seals. (Old Cars photo)

The President State series also included the limited-production Speedster model. It included many options, including those seen here, as standard equipment, but had a rather steep price tag. (Studebaker photo)

Even the President State four-door sedan was sporty. With a big V-8, top speed was more than 100 miles per hour. Road handling, cornering characteristics and suspension got very high marks. (Applegate & Applegate)

Rare, radical 1948 Tucker

By Staff

Engineer Preston Tucker of Ypsilanti, Michigan wanted to build a car that was safe, fast and comfortable. He got started on his project after World War II and the result was the rare and radical 1948 Tucker Torpedo.

Tucker collaborated on the design of his futuristic car with Alex Tremulis, renowned designer of the time who had been with Auburn Automobile Company, the manufacturer of Auburns, Cords and Duesenbergs. Originally, the Tucker was to be even more radical than the car that was eventually produced.

Initial plans called for a steering wheel in the center of the car and front fenders that turned in the direction in which the wheels were steered. Although both of these concepts were scrapped, a third headlamp in the middle of the front end was retained as a production line feature. It was called a "cyclop's eye" and rotated with the wheels as they were steered by the driver.

The Tucker's engine was located at the back of the car, between the rear wheels. It displaced 335 cubic inches and produced 166 horsepower. Both of these numbers were unusually high for the late-1940s, but then the Tucker was an unusual automobile. For instance, the engine was purposely positioned lower than the passenger seat, as it was felt that this would diminish heat, noise and fumes and keep all three from entering the car's interior.

Standing only 60 inches high off the roadway, the roof of the Tucker was also extremely low for cars of the early postwar era. Its wheelbase was 128 inches long ... somewhat large ... and it had a hefty weight factor of 4,235 pounds.

Preston Tucker worked hard to raise $25 million through stock sales. He purchased a factory in Chicago, Illinois which Chrysler Corporation had used during World War II to manufacture aircraft engines. Production of Tucker Torpedos started in 1946 and continued until 1948. However, it was a hand-to-mouth operation the entire duration and only a very limited number of cars were ever completed. There is some debate about the exact number, but the generally accepted total is 51 units.

Thirty-seven cars were completed and sold. The 38th was a prototype. All the rest were still on the assembly line when the plant was forced to close after the government stepped in. Tucker had been accused of stock manipulation and the factory was unable to continue operations.

The factory price for a Tucker in 1948 was $2,450. A dealer network was actually set up and showrooms around the country waited for cars to sell. Preston Tucker's problems with the Securities & Exchange Commission started early in the game, but in 1949 he was indicted.

Later, Tucker was cleared of the charges against him. However, it was 1949, not 1989, and his reputation in American industry was ruined. Later, he developed plans to build a small car in South America. Illness brought that effort to a sudden halt. Tucker died in 1956.

At least 49 Tuckers exist today. Among present-day Tucker owners is motion picture producer Francis Ford Coppola who, several years ago, created the film called: "Tucker: The Man and His Dream."

SPECIFICATIONS

Year 1948
Make .... Tucker
Model .... Torpedo
Body style .... four-door sedan
Base price $2,450
Engine .... horizontal-opposed; six-cylinder
Bore x stroke .... 4.50 x 3.50 inches
CID 334.1
Compression ratio 7.0:1
Carburetor 2V
H.P. .... 166 @ 3200 rpm
Wheelbase .... 128 inches
Overall length .... 219 inches
Weight .... 4,235 pounds
Tires .... 7.00 x 15
SCAC Value $300,000

Tucker got started on a new project after World War II and the result was his rare and radical 1948 Torpedo. The "cyclops" headlamp above the center grille turned in the direction that the wheels were steered. (Jim Bach photo)

The rear of the Tucker Torpedo looked like a 1940s conception of a rocket ship. A very limited number of cars was completed. There is some debate about the exact amount built, but the generally accepted total is 51 units.

Standing only 60 inches high, the roof of the Tucker was extremely low for cars of the early postwar era. This car is part of the Imperial Palace Auto Collection at the Imperial Palace Hotel and Casino in Las Vegas, Nevada.

The rear-mounted engine was purposely positioned lower than the passenger seat, as it was felt that this would diminish heat, noise and fumes and keep all three from entering the car's interior.

On your marque

American Marque Car Clubs

General Interest

Antique Automobile Club of America, 501 West Governor Road, Post Office Box 417, Hershey, PA 17033. William H. Smith, (717) 534-1910.

Classic Car Club of America, O'Hare Lake Office Plaza, 2300 East Devon Avenue, Suite 126, Des Plaines, IL 60018. (708) 390-0443.

Contemporary Historical Vehicle Association, 314 Alyssum Circle, Nipomo, CA 93444-9208. Bill Mirken, (805) 929-6071.

Horseless Carriage Club of America, 128 South Cypress Street, Orange, CA 92666-1314. (714) 538-HCCA.

International Society for Vehicle Preservation, Post Office Box 50046, Tucson, AZ 85703-1046. (602) 622-2201.

Milestone Car Society, Post Office Box 24612, Speedway, IN 46224. Jerry Flanary, (317) 356-4246.

Society of Automotive Historians, Post Office Box 339, Matamoras, PA 18336.

Veteran Motor Car Club of America, 9829 Mary Ellen Place Northeast, Albuquerque, NM 87111. Howard Zorn, (505) 296-4001.

AMC

AMC World Clubs, Incorporated, 7963 Depew Street, Arvada, CO 80003. Larry Mitchell, (303) 428-8760.

American Motors Owners Association, 6756 Cornell Street, Portage, MI 49002. Darryl Salisbury, (616) 323-0369.

American Motorsport International, 7963 Depew Street, Arvada, CO 80003-2527. Larry Mitchell, (303) 428-8760.

American Motors/Rambler Club, 2645 Ashton Road, Cleveland Heights, OH 44118. Frank Wrenick, (216) 371-5946.

The Classic AMX Club International, 7963 Depew Street, Arvada, CO 80003-2527. Larry Mitchell, (303) 428-8760.

Classic AMX Registry, 21 Creek Road, Dauphin, PA 17018. Lee Peterson, (717) 921-3363.

National American Motors Driver & Racers Association (NAMDRA), Post Office Box 987, Twin Lakes, WI 53181. Jock Jocewicz, (414) 396-9552.

Auburn/Cord/Duesenberg

Auburn-Cord-Duesenberg, Incorporated, 9N089 Corron Road, Elgin, IL 60123. J.P. Corbin, (708) 464-5767.

Austin

American Austin/Bantam Club, Route 1, Box 137 - 351 Wilson Road West, Willshire, OH 45898. Helen Jean White, (419) 495-2569.

Avanti

Avanti Owners Association, International, Post Office Box 28788, Dallas, TX 75228. 1 (800) 527-3452.

Buick

Buick Compact Club of America (1961-1962-1963 Specials & Skylarks), Route 1, Box 39B, Marion, TX 78124. (210) 625-5914.

Buick Club of America, Post Office Box 401927, Hesperia, CA 92340-1927. Val Ingram, (619) 947-2485.

Buick GS Club of America, 1213 Gornto Road, Valdosta, GA 31602. (912) 244-0577.

1950 Buick Registry, 54 Madison Street, Pequannock, NV 07440. Bill Braga, (201) 696-8418.

1953-1954 Buick Skylark Club, Post Office Box 1281, Frederick, MD 21702. Richard Beckley, Jr., (301) 898-5137.

Riviera Owners Association, Post Office Box 26344, Lakewood, CO 80226. Ray Knott, (303) 987-3712.

1937-1938 Buick Club, 1005 Rilma Lane, Los Altos, CA 94022. Harry Logan, (415) 941-4587.

Cadillac

Cadillac Drivers Club, 5825 Vista Avenue, Sacramento, CA 95824. Wray Tibbs, (916) 421-3193.

Cadillac LaSalle Club, Incorporated, 3083 Howard Road, Petoskey, MI 49770. (616) 347-4611.

1958 Cadillac Owners Association, Post Office Box 29, Braintree, MA 02184. Dave Becker, (617) 843-4485. (Antique Automobile Club of America)

Single Cylinder Cadillac Registry, 311 Nature Trail Drive, Greer, SC 29651-9054. Paul Ianvario, (803) 879-4587.

Checker

Checker Car Club of America, 469 Tremaine Avenue, Kenmore, NY 14217. Donald McHenry, (716) 877-3358.

Chevrolet

Bow Tie Chevy Association, Post Office Box 608108, Orlando, FL 32860. Denny Williams, (407) 880-1956.

Chevrolet Nomad Association, 2537 South 87th Avenue, Omaha, NE 68124. Bob Maline, (402) 393-7281.

Chevy Performance Club of America, Post Office Box 4306, South Bend, IN 46634. Bruce Bukland, (219) 232-0085.

Corvair Society of America (CORSA), Post Office Box 607, Lemont, IL 60439-0607. Harry Jensen, (708) 257-6530.

Corvette Club of America, Post Office Box 9879, Bowling Green, KY 42102-9879. Keith Blandford, (502) 451-5680.

Corvettes Limited, Incorporated, 11 Liberty Ridge Trail, Totowa, NJ 07512. Neal Ventola, (201) 338-4763.

Corvette ZR-1 Registry, 29 Lucille Drive, Sayville, NY 11782.

Cosworth Vega Owners Association, Post Office Box 1783, Bloomington, IN 47402. Bob Chin, (812) 339-0838.

International Camaro Club, Incorporated, 2001 Pittston Avenue, Scranton, PA 18505-3233, (717) 347-5839.

Late Great Chevys, Post Office Box 607824, Orlando, FL 32860. (407) 886-1963.

National Association of Chevrolet Owners, Post Office Box 9879, Bowling Green, KY 42102-9879. Keith Blandford, (502) 451-5680.

National Chevelle Owners Association, 7343-J West Friendly Avenue, Greensboro, NC 27410, (919) 854-8935.

The National Chevy Association (1953-1954 Chevys), 947 Arcade, Saint Paul, MN 55106. (612) 778-9522.

National Corvette Owners Association, 900 South Washington Street, Falls Church, VA 22046. (703) 533-7222.

National Corvette Restorers Society, 6291 Day Road, Cincinnati, OH 45252-1334. Gary Mortimer, (513) 385-8526.

National Impala Association, Post Office Box 968, Spearfish, SD 57783. (605) 642-5864.

National Monte Carlo Owners Association, Post Office Box 187, Independence, KY 41051. (606) 491-2378.

National Nostalgic Nova, Post Office Box 2344, York, PA 17405. (717) 252-4192.

North-East Chevy/GMC Truck Club, Post Office Box 155, Millers Falls, MA 01349. Mike DeWick, (508) 371-7477.

Obsolete Fleet Chevys (1955-1956-1957 Chevys), Post Office Box 554, McMinnville, OR 97128. Jerry Kwiatkowski, (503) 472-4382.

1965-1966 Full Size Chevrolet Club, 15615 Street Road 23, Granger, IN 46530. Harold Foos, (219) 272-6964.

Straight-Axle Corvette Enthusiasts, Post Office Box 2288, North Highlands, CA 95660. (916) 729-1165.

Suburban Driver Club (Chevy/GMC Suburbans), Post Office Box 292, Orchard Park, NY 14127.

Tri-Chevy Association, 24862 Ridge Road, Elwood, IL 60421. (815) 478-3633.

U.S. Camaro Club, Post Office Box 608167, Orlando, FL 32860. (407) 880-1967.

Vintage Chevrolet Club of America, Post Office Box 5387, Orange, CA 92613-5387. Shirley Whitesell, (714) 633-1310.

Chrysler/DeSoto/Dodge/Plymouth

Airflow Club of America, 609 Crown Colony Drive, Arlington, TX 76006. Tony Palmer, (817) 261-1334.

Chrysler 300 Club International, 4900 Jonesville Road, Jonesville, MI 49250. Eleanor Riehl, (517) 849-2783.

Chrysler Maserati TC Registry, Post Office Box 66813, Chicago, IL 60666-9998.

Chrysler Product Owners Club, Incorporated, 806 Winhall Way, Silver Spring, MD 20904. Ray Montgomery, (301) 622-2962.

Chrysler Town & Country Owners Registry, 406 West 34th, Kansas City, MO 64111. Dennis McLaughlin, (816) 931-3341.

D.A.R.T.S. (1967-1972 Dodge Darts & Demons), Post Office Box 9, Wethersfield, CT 06129-0009.

Daytona-Superbird Auto Club, 13717 West Green Meadow Drive, New Berlin, WI 53151. (414) 786-8413.

DeSoto Club of America, 105 East 96th, Kansas City, MO 64114. Walter O'Kelly, (816) 421-6006.

Dodge Brothers Club, Incorporated, 4451 Wise Road, Freeland, MI 48623. Tom Mills, (412) 654-2893.

Early Hemi Association, 233 Rogue River Highway, Suite 354, Grants Pass, OR 97527-5477. (503) 476-7422.

Hurst 300 Registry, 5844 West Eddy Street, Chicago, IL 60634. Roman Robaszewski, (312) 685-4980.

LRT/Warlock Registry, Post Office Box 578, Highland City, FL 33846.

Mopar Rapid Transit Club, 10705 Old Beatly Ford, Rockwell, NC 28138. Lee Foster, (704) 279-7410.

National DeSoto Club, Incorporated, 412 Cumnock Road, Inverness, IL 60067.

National Hemi Owners Association, 1693 South Reese Road, Reese, MI 48757. Marc McKenney, (517) 868-4921.

Plymouth Owners Club, Incorporated, Post Office Box 416, Cavalier, ND 58220. Jim Benjaminson, (701) 549-3746.

Slant 6 Club of America, Post Office Box 4414, Salem, OR 97302. Jack Poehler, (503) 581-2230.

WPC Club, Incorporated (Chrysler Products), Post Office Box 3504, Kalamazoo, MI 49003-3504. Richard Bowman, (616) 375-5535.

Crosley

Crosley Automobile Club, 217 North Gilbert, Iowa City, IA 52245. Jim Friday, (319) 338-9132.

Durant

Durant Family Registry, 2700 Timber Lane, Green Bay, WI 54303. Jeff Gillis, (414) 499-8797.

Ford/Lincoln/Mercury/Edsel

Big M Mercury Club, 5 Robinson Road, West Woburn, MA 01801.

Capri Car Club, Limited, Post Office Box 111221, Aurora, CO 80042-1221. Tom Johnston, (303) 343-8604.

Classic Thunderbird Club International, Post Office Box 4148, Santa Fe Springs, CA 90670-1148. Margie Price, (310) 945-6836.

Cobra Jet Registry, 6890 Plainfield Road, Dearborn, MI 48127.

Cougar Club of America, O-4211 North 120th Avenue, Holland, MI 49424. John Baumann (616) 396-0390.

Crown Victoria Association, Post Office Box 6, Bryan, OH 43506. Robert Haas, (313) 248-3400.

Cyclone Spoiler/Talladega Registry, Post Office Box 422, Alpharetta, GA 30239-0422.

Early Ford V-8 Club of America, Post Office Box 2122, San Leandro, CA 94577. Jerry Windle, (619) 283-8117.

Econo (Econoline vans), 15039 Costela Street, San Leandro, CA 94579.

Edsel Owners Club, Incorporated, 4713 Queal Drive, Shawnee, KS 66203.

Fabulous Fifties Ford Club of America, Post Office Box 286, Riverside, CA 92502. (714) 354-5667.

Fairlane Club of America, 2116 Manville Road, Muncie, IN 47302-4854.

Falcon Club of America, Post Office Box 113, Jacksonville, AR 72076.

FoMoCo Owners Club, Post Office Box 19665, Denver, CO 80219. Barry L. Abels, (303) 935-6662.

Ford and Mercury Restorers Club, Post Office Box 2133, Dearborn, MI 48123. Robert Haas, (313) 248-3400.

Ford F-100 Truck Club, 1315 Hollis Terrace, Bremerton, WA 98310. John Ramage, (206) 377-9828.

Ford Galaxie Club of America, Post Office Box 2206, Bremerton, WA 98310. William Barber, (206) 377-4957.

1949-1950-1951 Ford/Mercury Owners, Post Office Box 30647, Midwest City, OK 73140-3647. Mike McCarville, (405) 737-6021.

1949-1959 Ford Owners Association, 12204 43rd Avenue South, Tukwila, WA 98178. Cheryl Matson, (206) 762-1379.

1954 Ford Club of America, 1517 North Wilmot, Number 144, Tucson, AZ 85712. John Strobeck, (602) 886-1184.

Heartland Vintage Thunderbird Club (1958-1969), 5002 Gardner, Kansas City, MO 64120. Don Kimrey, (816) 353-6151.

International Edsel Club, Post Office Box 371, Sully, IA 50251. (515) 594-4284.

International Mercury Owners Association, 6445 West Grand Avenue, Chicago, IL 60635. Jerry Robbin, (312) 622-6445.

Lincoln & Continental Owners Club, Post Office Box 68308, Portland, OR 97268. Becky D'Ambrosia, (503) 659-3769.

Lincoln Owners Club, 22 Spring Street, Cary, IL 60013. Jim Griffin, (715) 356-3039.

Lincoln Zephyr Owners Club, Post Office Box 165835, Miami, FL 33116. John Murphy, (305) 274-3624.

Mid Century Mercury Car Club, 4411 Harrison Road, Kenosha, WI 53142.

Model A Ford Cabriolet Club, Post Office Box 515, Porter, TX 77365. Larry Machacek, (713) 429-2505.

Model A Ford Club of America, 250 South Cypress, La Habra, CA 90631. Jerry E. Wilhelm, (310) 697-2712.

Model A Ford Foundation, Incorporated, 302 Pine Shadow Lane, Suite 500, Lake Mary, FL 32746. Gary Kosen, (407) 323-6885.

Model A Restorers Club, 24800 Michigan Avenue, Dearborn, MI 48124-1713. Steve Smith, (313) 326-2428.

Model T Ford Club International, Post Office Box 438315, Chicago, IL 60643-8315.

Model T Ford Club of America, Post Office Box 743936, Dallas, TX 75374-3936. Barbara Klehfoth, (214) 783-7531.

Mustang Club of America, Post Office Box 447, Lithonia, GA 30058-0447. Bill Koivu, (404) 482-4822.

Mustang Owners Club International, 2720 Tennessee North East, Albuquerque, NM 87110. Paul McLaughlin, (505) 296-2554.

Pantera International, 18586 Main Street, Suite 100, Huntington Beach, CA 92648-1720. (714) 848-6674.

Performance Ford Club of America, 13155 U.S. Route 23, Ashville, OH 44907. (614) 983-2273.

Ranchero Club, 1339 Beverly Road, Port Vue, PA 15133.

Road Race Lincoln Register, 461 Woodland Drive, Wisconsin Rapids, WI 54494. Burr Oxley, (715) 423-9579.

Shelby American Automobile Club, Post Office Box 788, Sharon, CT 06069.

Shelby Owners of America, Incorporated, Post Office Drawer 1429, Great Bend, KS 67530.

Special Interest Fords of the 1950s, 246 Silvercreek, Duncanville, TX 75137. Jack Baird, (214) 298-4797.

Thunderbirds of America, Post Office Box 2766, Cedar Rapids, IA 52406. John Draxler, (712) 884-6546.

Vintage Thunderbird Club International, Post Office Box 2250, Dearborn, MI 48123-2250. Maurice E. Black, (817) 497-3816.

Franklin

The H.H. Franklin Club, Cazenovia College, Cazenovia, NY 13035.

Gardner

Gardner Automobile Register, 341 Fitch Hill Road, Uncasville, CT 06382. Ed Jacobowitz, (203) 848-8934.

Graham

Graham Owners Club, Incorporated, 2909 13th Street, Wausau, WI 54401. Mike Keller, (715) 845-1507.

Hudson

Hudson-Essex-Terraplane Club, Incorporated, Post Office Box 215, Milford, IN 46542. Paul Schuster, (412) 462-2058.

Hupmobile

Hupmobile Club, Incorporated, 158 Pond Road, North Franklin, CT 06154. Steve Christie, (203) 642-6697.

Jordan

Jordan Register, 5231 Stratford Avenue, Westminster, CA 92683.

Kaiser/Frazer

Kaiser-Darrin Owners Roster, RD 3, Box 36, Somerset, PA 15501. Dave Antram, (814) 445-6135.
Kaiser-Frazer Owners Club, Incorporated, Post Office Box 1251, Wellsville, NY 14895. Dave Choate, (716) 593-4751.

Kissel

Kissel Kar Klub, 147 North Rural Street, Hartford, WI 53027. Dale Anderson, (414) 673-7999.

King Midget

King Midget Club, Post Office Box 549, Westport, IN 47283.
King Midget International Car Club, 385 Cavan Drive, Pittsburgh, PA 15236.

Locomobile

Locomobile Society of America, 3165 California Street, San Francisco, CA 94115-2412, (415) 563-1771.

Marmon

Marmon Club, Post Office Box 8031, Canton, OH 44711. Bruce Williams, (216) 454-7070.

Maxwell

Maxwell Registry, RD 4, Box 8, Ligonier, PA 15658. Tom Thoburn, (412) 238-6397.
The Maxwell-Briscoe Registry, 55 East Golden Lake Road, Circle Pines, MN 55014. James Moe, (612) 786-6609.

Mercer

Mercer Associates, 414 Lincoln Ave., Havertown, PA 19083. John Rendemonti, (215) 446-0138.

Metz

Metz Register, 721 East State Street, Millsboro, DE 19966. Don Addor, (302) 934-8021.

Mitchell

Mitchell Roster, 695 U.S. Highway 6, Coal Valley, IL 61240. Don Mitchell, (309) 799-5263.

Nash/Metropolitan

Charles W. Nash Association, 2412 Lincoln Avenue, Alameda, CA 94501.
Metropolitan Owners Club of North America, 5009 Barton Road, Madison, WI 53711. Larry Hurley, (608) 271-0457.
Nash Car Club of America, 1-North-274 Prairie, Glen Ellyn, IL 60137.
1935-1936 Nash Registry, 2412 Lincoln Avenue, Alameda, CA 94501. Sieg Wroebel (415) 523-0454.

Oldsmobile

Curved Dash Oldsmobile Club, 3455 Florida Avenue North, Minneapolis, MN 55427. Gary Hoonsbeen, (612) 533-4280.
Hurst/Olds Club of America, 1600 Knight Road, Ann Arbor, MI 48103-9303. (313) 994-8778.
Oldsmobile Club of America, Post Office Box 16216, Lansing, MI 48901. Dennis Casteele, (517) 321-8825.
Oldsmobile Club of America-1957 Chapter, Post Office Box 3712, Arcadia, CA 91066.
National Antique Oldsmobile Club, Post Office Box 915, Fremont, OH 43420-0915.

Packard

Packard Automobile Classics, 420 South Ludlow Street, Dayton, OH 45420. 1 (800) 527-3452.
Packards International Motor Car Club, 302 French Street, Santa Ana, CA 92701. (714) 541-8431.
The Packard Club, Post Office Box 2808, Oakland, CA 94618. Stella Pyrtek-Blond, (908) 738-7859.
1937 Packard Six International Roster, 3174 White Tail Lane, Adel, IA 50003-9724. Kevin Rice, (515) 993-4456.

Paige

Paige Registry, Post Office Box 66, Shady, NY 12409.

Pierce-Arrow

Pierce-Arrow Society, Incorporated, 135 Edgerton Street, Rochester, NY 14607.

Pontiac/Oakland

All American Oakland Chapter, 22 Washington Street, Millinocket, ME 04462. John Armstrong, (614) 878-9536.
Fiero Owners Club of America, 215 North State College, Orange, CA 92668. Phil Huff, (714) 978-3132.
GTO Association of America, 1634 Briarson Drive, Saginaw, MI 48603.
The Judge GTO International, 114 Prince George Drive, Hampton, VA 23669. Robert J. McKenzie, (804) 838-2059.
Oakland/Pontiac Enthusiasts Organization, Incorporated, Post Office Box 0371, Drayton Plains, MI 48330. (313) 623-7573.
Original GTO Club, Post Office Box 18438, Milwaukee, WI 53218. Neil Moderson, (414) 691-2627.
Pontiac-Oakland Club International, Incorporated, 286 Ahmu Terrace, Vista, CA 92084. Dick Hoyt, 1 (800) 457-POCI.

Reo

Reo Club of America, Incorporated, Post Office Box 336, Sea Bright, NJ 07760.

Saxon

Saxon Registry, 5250 North West Highland Drive, Corvallis, OR 97330. Walter Prichard, (503) 752-6231.

Studebaker

Antique Studebaker Club, Post Office Box 28845, Dallas, TX 75228-0845. Sheldon Harrison, 1 (800) 527-3452.
1953-1954 Studebaker Coupe/Hardtop Owners, 3540 Middlefield Road, Menlo Park, CA 94025. Dennis Hommel, (415) 365-4565.
1956 Studebaker Golden Hawk Owners Club, 1025 Nodding Pines Way, Casselberry, FL 32707. Frank J. Ambrogio, (407) 699-8446.
Studebaker Drivers Club, Post Office Box 28788, Dallas, TX 75228-0788. Sheldon Harrison 1 (800) 527-3452.

Stutz

Stutz Club, Incorporated, 7400 Lantern Road, Indianapolis, IN 46256. Bill Greer, (317) 849-3443.

Tucker

Tucker Automobile Club of America, 311 West 18th Street, Tifton, GA 31794. Bill Wells, (912) 382-4573.

Whippet

Model 96 Whippet Newsletter, RR3, Box 28-A, Parsons, KS 67357. John Olson, (316) 421-0643.

Willys/Overland/Knight

Willys Aero Survival Count, 952 Ashbury Heights Court, Decatur, GA 30030-4177. Rick Kamen, (404) 288-8222.
Willys Club, 719 Lehigh Street, Bowmanstown, PA 18030. (215) 852-3110.
Willys-Overland-Knight Registry, Incorporated, 1440 Woodacre Drive, McLean, VA 22101-2535.